INDUSTRIAL GALACTOMANNAN POLYSACCHARIDES

T0239886

INDUSTRIAL GALACTOMANNAN POLYSACCHARIDES

N. K. MATHUR

CRC Press
Taylor & Francis Group
Boca Raton London New York

CRC Press is an imprint of the
Taylor & Francis Group, an **informa** business

CRC Press
Taylor & Francis Group
6000 Broken Sound Parkway NW, Suite 300
Boca Raton, FL 33487-2742

First issued in paperback 2017

© 2012 by Taylor & Francis Group, LLC
CRC Press is an imprint of Taylor & Francis Group, an Informa business

No claim to original U.S. Government works

ISBN 13: 978-1-138-11478-4 (pbk)
ISBN 13: 978-1-4398-4628-5 (hbk)

This book contains information obtained from authentic and highly regarded sources. Reasonable efforts have been made to publish reliable data and information, but the author and publisher cannot assume responsibility for the validity of all materials or the consequences of their use. The authors and publishers have attempted to trace the copyright holders of all material reproduced in this publication and apologize to copyright holders if permission to publish in this form has not been obtained. If any copyright material has not been acknowledged please write and let us know so we may rectify in any future reprint.

Except as permitted under U.S. Copyright Law, no part of this book may be reprinted, reproduced, transmitted, or utilized in any form by any electronic, mechanical, or other means, now known or hereafter invented, including photocopying, microfilming, and recording, or in any information storage or retrieval system, without written permission from the publishers.

For permission to photocopy or use material electronically from this work, please access www.copyright.com (http://www.copyright.com/) or contact the Copyright Clearance Center, Inc. (CCC), 222 Rosewood Drive, Danvers, MA 01923, 978-750-8400. CCC is a not-for-profit organization that provides licenses and registration for a variety of users. For organizations that have been granted a photocopy license by the CCC, a separate system of payment has been arranged.

Trademark Notice: Product or corporate names may be trademarks or registered trademarks, and are used only for identification and explanation without intent to infringe.

Visit the Taylor & Francis Web site at
http://www.taylorandfrancis.com

and the CRC Press Web site at
http://www.crcpress.com

While studying chemistry at the undergraduate level, I was very much impressed by Herman Emil Fisher's research on carbohydrates and the logical conclusions drawn from it, which resulted in the elucidation of the chemical structure, including the configuration of glucose and other sugars. This created in me, an everlasting interest in the chemistry of carbohydrates.

My interest in carbohydrates, including their vast industrial uses was later rekindled in 1970 with my association with the guar gum (a galactomannan polysaccharide) industry in my hometown of Jodhpur, India.

The late Dr. H. C. Srivastava, senior joint director, Ahmedabad Textile Industry Research Association, Ahmedabad, was a leading carbohydrate chemist in India. I had known him only through correspondence, until I met him at the Central Food Technological Research Institute at Mysore (India) during the third annual conference in 1988 of the Association of Carbohydrate Chemists and Technologists (India).

Dr. Srivastava told me about his plan of writing a book on galactomannans and knowing about my interest in this specialized field and my long association with the guar gum industry, he asked me to give him relevant literature on galactomannan gums, which I had collected on galactomannan polysaccharides. Unfortunately, his plan of writing a monograph on galactomannan polysaccharides did not materialize due to his untimely death in 1995.

I have now compiled this book on industrial galactomannan polysaccharides, and I consider this as my humble tribute to Dr. Srivastava, who had dreamt of writing such a book.

I dedicate this book to the memory and inspiration for writing it, which I got from Dr. H. C. Srivastava.

It is for the readers to judge how far I have succeeded in this objective.

Contents

Preface

I have been associated with the guar gum (an industrial galactomannan polysaccharide) industry at Jodhpur, which is a city in the northwestern Rajasthan state of India, for nearly four decades now. Guar seed processing into its gum was started in India in late 1950s and early 1960s. This was based on the transfer of technology by two U.S. companies (General Mills Co. and Stein, Hall & Co.) and establishing guar galactomannan-based industries in India with their collaboration. Subsequently similar industries were also established in Pakistan. With the establishment of these industries on the Indian subcontinent, processing of guar seed into gum powder was gradually reduced in the United States and Europe. However, the export of guar seed split from India has continued even now to most of Europe, Japan, and the United States, where some value-added specialty guar gum products are still manufactured according to secret and patented procedures. During the past two decades, there has been an increasing trend of outsourcing production of these so-called specialty products to Indian manufacturers.

For several years the export-oriented guar gum industry in India flourished and progressed well. This was mainly due to mastering of the technology of mechanical processing of guar seed into gum powder, without much understanding of underlying science and technology of the process and the product. India currently produces over 70% of the world production of guar seeds, most of which is processed into guar gum products and its by-products, including protein-rich cattle feed. Many guar gum industries in India are now emerging as multihydrocolloid industries processing, importing, or trading in hydrocolloids other than guar gum.

It was in the late 1960s when I returned from my sabbatical leave from McGill University (Montreal, Canada) that I was approached by the guar gum industry at Jodhpur and certain Rajasthan state agencies (industry department) to take an interest in the development of this industry. Since then, I have been closely associated with this industry and its developments, including training and human resource development. This has provided me with an opportunity for an overall study of the science and technology of processing and quality control of guar gum and other galactomannan polysaccharides and their applications.

In my academic pursuit, I have carried extensive research, both pure and applied, on galactomannans. Besides guar gum, I have carried out an extensive study of other galactomannan and polysaccharides. As a result of this, I have attempted to write this book.

Plant seeds are a source of many hydrocolloids, particularly polysaccharide gums. Most edible seeds contain starch as their reserve polysaccharide, but a large number of legume seed endosperms from annual agriculture legume crops as well as from some perennial legume trees have galactomannans as their reserve polysaccharides. Seed galactomannans, for example, guar gum and locust bean gum, are water dispersible, high molecular weight polysaccharides of colloidal dimension.

These are also known as *plant seed hydrocolloids* or *gums*. These terms have been used interchangeably in this book and elsewhere in literature related to these polysaccharides.

The biological role of galactomannans in legume seeds is not very clear, and it has been a subject of speculation by scientists. Being the reserve polysaccharide in legume seeds, galactomannans serve as a source of carbon for their germinating seedlings. Only after emerging from the soil, the plant starts photosynthesis using atmospheric carbon dioxide. Seed galactomannans also guard against water stress during water scarcity periods. They control and manipulate water absorption by the seeds of different legume species and create a favorable condition for their growth. Man, to his own advantages, has utilized this unique water-holding property of galactomannans. Thus, these polysaccharides are being used as additives for food and as general-purpose hydrocolloids for many other industries.

A number of galactomannans from different plant sources are being produced commercially, and these have become industrial commodities. Each one of these galactomannans has some characteristic functional properties, which are determined by its molecular weight, the ratio of mannose to galactose, and their molecular architecture or chemical structure. Hence, one galactomannan may not always be replaceable by another one in a specific application. This has been the reason why so many galactomannans have to be undertaken for commercial production. Structural variations and occurrence of a broad spectrum of these plant polysaccharides have been a windfall for the consumer industry and provided a large range of applications based on their well-defined structures and functionalities. Different galactomannans have variable availability and price differences. There is a large import–export and trading of galactomannan gums and other hydrocolloids at the international level.

Until now, a comprehensive compilation of the information related to galactomannans in the form of an exclusive book has not been published. Such a monograph could be useful to scientists and technologists in the galactomannan industry as well as to the end users of galactomannans in food and nonfood industries, and to a broad range of carbohydrate scientists and academicians. Generally, the topic of galactomannan polysaccharides is dealt only as a chapter in books on carbohydrates, polysaccharides, and gums. Exclusive monographs have been written on certain other polysaccharides, including cellulose, starch, tree exudates, and microbial polysaccharides. This is the first monograph to be written on industrial galactomannan polysaccharides, which have now acquired importance close to cellulose gums and starches.

The book aims at compiling information about industrial galactomannans, which should be useful to the manufacturers, traders, and end users of galactomannans, technologists in the polysaccharides-related industry, and scientists and academics interested in carbohydrates.

Locust bean gum is the oldest, commercially produced galactomannan. Guar gum and few other galactomannans are now commercially produced, about which detailed information has been included in this book. These other galactomannans, included herein, are tara gum, fenugreek gum, *Cassia tora* gum, and *Cassia fistula* gum.

Glucomannan polysaccharides, which bear some similarities to galactomannans, have also been briefly referred.

Cassia tora and *Sesbania bispinosa* gums are two galactomannans that are derived from wild, annual herbs. *Cassia fistula* gum is yet another galactomannan polysaccharide, derived from a tree, which as yet is not produced commercially, but it has strong potential for commercialization. *Cassia fistula* is a widely occurring tropical tree of Southeast Asia, which produces a unique galactomannan polysaccharide in its seed. The tara shrub from Peru has become a source of industrial galactomannan gum, and it is now the time that the *Cassia fistula* tree gets recognized as yet another tree that can become a source of industrial gum extraction. I have strongly pleaded for commercialization of *Cassia fistula* gum, particularly looking to its desirable functional properties. It is my sincere hope that looking to the demand for locust bean gum–type galactomannans, the gum industry in India will soon consider starting commercial production of *Cassia fistula* gum also. A general reference has also been made to other galactomannan-bearing plants, which can become a future source of industrial galactomannan gums.

Six general chapters have been included in this monograph, in addition to the chapters on individual galactomannan polysaccharides. These general chapters are

Chapter 1: General Introduction to Carbohydrate
Chapter 2: Galactomannan Polysaccharides
Chapter 3: Hydrocolloids or Gums
Chapter 4: Interactions of Galactomannans
Chapter 5: Rheology of Hydrocolloids
Chapter 6: Derivatization of Polysaccharides

The information included in these general chapters should be helpful for scientists and technologists in the polysaccharide industry as well as to the research scientists and academicians concerned with research and development on galactomannans or other hydrocolloids. They can get the required information related to polysaccharide hydrocolloids from this single source. References given in each chapter have been balanced, so as to provide help to any reader to consult further literature on polysaccharides in general.

In the chapters on individual galactomannans, a brief history of the product gum, cultivation of the concerned plant source and the area of its cultivation, brief description of the plant and its seed, manufacturing process, chemical structure, functional properties uses, and applications of various galactomannans are described. More details of these are given in chapters on guar gum (Chapter 7) and locust bean Gum (Chapter 8), which are the most representative galactomannans. In some places, some descriptions may look repetitive, but since a reader may only have selective interest for specific information related to a particular galactomannan, inclusion of such information at appropriate places was deemed necessary.

Chapters on the rheology of hydrocolloids (Chapter 5), interaction of galactomannans (Chapter 4), and derivatization of polysaccharides (Chapter 6) are more technical and based on scientific principles, and these may be somewhat difficult to grasp by persons lacking a scientific background in carbohydrates. Frequent reference to these chapters should be of help to such persons in various fields of the galactomannan industry.

I sincerely hope that this comprehensive compilation of information on galactomannan gums in a book form is useful to concerned persons, and I welcome any suggestions and additional information on this subject.

Acknowledgments

Dr. C. K. Narang, my former colleague and research collaborator, at Jai Narain Vyas University, Jodhpur (India); Dr. K. C. Gupta, Director, Central Toxicological Research Institute (CSIR) at Lucknow (India); and Dr. B. P. Nagori, Director, Pharmacy, L. M. College of Science and Technology, Jodhpur, are among my several former research students who worked on guar, galactomannan polysaccharides, and related topics. I am thankful to them and all my other research collaborators who did research on guar gum and other polysaccharides in my group. Guar gum industrial units in Jodhpur, namely, Lucid Colloids, Shriram Gum, and the Sunita Minechem Industry gave me a free hand in studying their manufacturing processes, quality control systems, and research and development activities related to the guar industry, which helped me in improving my grasp of galactomannan chemistry and industrial technology. I acknowledge my sincere thanks to all of them.

Author

N. K. Mathur was born in 1930 in Parbatsar, a remote village in the state of Rajasthan, India. He was educated in the city of Jodhpur and, in addition to extensive teaching and research on a variety of subjects at Jai Narain Vyas University (Jodhpur), he has done research at McGill University Montreal (Canada) and Ottawa at the National Research Council of Canada laboratories.

Mathur has more than 125 research publications in varied fields, including two books of international repute published by Academic Press. He has traveled extensively in Europe, Canada, United States, Israel, Egypt, and China and lectured at various institutions.

After his retirement from the university at Jodhpur, Mathur continued to work as a consultant in galactomannan (guar gum)-related industries. He carried out extensive research and development work in the field of industrial polysaccharides. Being from Jodhpur, which is the largest center for processing guar and other galactomannan polysaccharides, Mathur has expertise in the field of processing, applications, and research and development in the field of polysaccharides.

1 General Introduction to Carbohydrates

1.1 CARBOHYDRATES, THE PRIMARY PRODUCTS OF PHOTOSYNTHESIS

Carbohydrates are the most abundant, natural organic compounds. They are synthesized and consumed by nearly all plants, animals, and microorganisms of our planet. In animals, including man, glucose oxidation to carbon dioxide, provides the metabolic energy. Plants and animals store energy molecules in the form of *reserve carbohydrate polymers*. These include starches in many edible seeds, kernel polysaccharide in tamarind seeds, galactomannan polysaccharides in many legume plant seeds, and liver glycogen in man and animals. Plants also synthesize cellulose and hemicellulose to construct their structural framework and cell wall.

Carbohydrates have played a major role in the evolution of life, and advancement of civilization and agriculture on earth. They are essential for all human activities, providing fuel and food, and for overall development. Carbohydrates are one of the *macronutrients* and provide energy to the human body. In most cases, around four calories of energy is produced per gram of metabolized carbohydrate.

Carbohydrates are used by man, not only as food, but also in the form of clothes (cotton) he wears, furniture he uses, building materials for his habitat, and paper used in writing and for printing of books and currency notes. Besides serving as an energy source for man and structural component for plants, carbohydrates are also the molecular components of nucleotides, which carry the genetic information of all living beings. When conjugated to a variety of lipophylic biomolecules, for example steroids, in the form of glycosides, carbohydrates make them water dispersible and transportable from one part of the body to another part.

1.2 RELATIVE ABUNDANCE OF COMMON CARBOHYDRATES

Common and naturally occurring monosaccharide hexoses are D-glucose, D-fructose, D-mannose, and D-galactose. Common pentoses are D-xylose and L-arabinose, and in a special sense, D-ribose and 2-deoxy-ribose are ubiquitous because they are the components of RNA and DNA, which are present in all living cells, as the components of nucleic acids. Disaccharide sucrose, along with its component monosaccharide, is present in all fruits, sugar cane, and honey. Milk sugar (lactose) has D-galactose and D-glucose as its monosaccharide components. Many oligosaccharides and polysaccharides have variable sugars as their components.

Glucose (derived from the Greek word *gly* meaning "sweet") is present in human blood and under abnormal conditions in urine. Glucose is the exclusive sugar component of certain polysaccharides (e.g., cellulose, starch, and glycogen), and it is commercially produced from corn and potato starch. Cellulose is the most abundant material in biomass.

1.3   CLASSIFICATION OF CARBOHYDRATES

The name *carbohydrates* is derived from the fact that most of them can generally be represented by a chemical formula, $C_n(H_2O)_m$, that is, as the hydrates of carbon, exceptions being the deoxy sugars. The simplest carbohydrates—the monosaccharides—are linear, polyhydroxy aldehydes, or polyhydroxy ketones. Monosaccharide having five or six carbon atoms generally exists as an internal, cyclic hemiacetal, which can have a five-membered furanose or a six-membered pyranose structures. Presence of an aldehyde or α-hydroxy-ketone group confers reducing properties to sugars, toward Fehling and other copper (II) reagents. Linear polysaccharides, having hundreds of monosaccharide units, and linked via glycoside bonds have only one, terminal reducing group and do not show reducing properties. Two or more monosaccharide molecules can join via glycoside (acetal) linkage to form other carbohydrates, including the polysaccharides.

An amino group is an additional functional group present in amino sugars (e.g., glucosamine), which is a component of many biomolecules (e.g., chitin).

In its chair form of pyranose structure, the most common sugar, glucose, has all the bulky hydroxyl groups in an equatorial configuration. The C-1 glycoside bond can assume α- (axial) or β- (equatorial) configurations. Configurational isomers of glucose—mannose and galactose—differ in having, respectively, the C-2 and C-3 hydroxyl in axial positions. Polysaccharides having mannose and galactose as their monosaccharide groups and differ from polymers of glucose because of these configurational differences.

Homopolysaccharides and heteropolysaccharides are named after their constituent sugars. Thus, galactomannans are composed of mannose (major component) and galactose (minor component) sugars.

1.4   PHOTOSYNTHESIS OF CARBOHYDRATES

An essential natural process taking place in the biosphere we live in is the process of *photosynthesis* in plants. This is the process that has permitted the evolution of life on earth as it exits now. Photosynthesis in chlorophyll-bearing green plants involves conversion of the atmospheric carbon dioxide and soil water into carbohydrates. The process of photosynthesis uses the catalytic activity of green, chlorophyll pigment and the sun's radiation energy to produce carbohydrates, and it is accompanied with simultaneous release of molecular oxygen into the atmosphere. The net chemical process, taking place during photosynthesis is represented by the following simple equation:

$$6CO_2 + 6H_2O \text{ (Chlorophyll, hv)} \rightarrow C_6H_{12}O_6 + 6O_2$$

All green plants apparently photosynthesize by a common route. An intermediate product of photosynthesis is 2-phosphoglyceric acid—$HOOC.CHOH.CH_2OPO(OH)_2$— which is then transformed via glucose into starch or the polymeric cell-wall components, for example, cellulose. Depending upon a plant type, the other polymeric products formed include hemicellulose, pectin, a variety of seaweed polysaccharides, and varying amounts of sucrose and other sugars. A variety of other homopolysaccharides and heteropolysaccharides are also produced. These include galactomannans, glucomannans, and exudate gums.

The carbohydrates thus produced are the storable form of energy in plants. Animals, directly or indirectly, utilize carbohydrates as food in the metabolic processes for their growth and the mechanical work they need to perform. When an animal (or human) metabolizes carbohydrates aerobically to generate energy, the net reaction is just the reverse of the photosynthesis, that is,

$$C_6H_{12}O_6 + 6O_2 \text{ (Respiration)} \rightarrow 6CO_2 + 6H_2O + \text{Energy}$$

Thus, there is a symbiotic relationship between the plants and the animals. Glucose, the primary products of plant photosynthesis in the plants, is largely converted into its polymeric products, that is, cellulose and starches. Of these two polysaccharides, cellulose is the structural polysaccharide in plants and it helps in growth and building up of the plant structure. Starch and galactomannans are two examples of *reserve polysaccharide* in plants, which, along with lipids, proteins, DNA, and several other constituents, accumulates in the seeds or the tubers in some cases (e.g., starch in potatoes and glucomannan in konjac tuber). Plant seeds and tubers serve as food for animals and humans, and a small portion of these seeds is used for regrowing new plants. When the plant grows, the reserve polysaccharide in a seedling is used up as a carbon source by the germinating plant embryo until it emerges out of the soil and it can start photosynthesis, utilizing atmospheric carbon dioxide.

1.5   CONFORMATION OF POLYSACCHARIDES

Linear oligosaccharides and polysaccharides can be considered to have been formed by successive linking of monosaccharide units at the glycoside carbon atom. When the successive sugar monomers are linked axially (α-glycoside linkage), there is a bending at each glycoside bond. This can result in the formation of a helical conformation, such as that in the amylose fraction of starch. In contrast to this, bis-equatorial (β-glycoside linkage) linking of successive sugar monomers produces a nearly straight, linear chain, for example, that in cellulose and most of the galactomannan molecules.

Being linear and due to extensive interchain and intrachain hydrogen bonding, bundles of cellulose molecules acquire a fibrous character. In case of the bis-equatorially linked mannan backbone of galactomannans, chain–chain hydrogen bonding is considerably reduced due to the presence of numerous, single galactose grafts. Hence, in a solution they can disperse to have a folded or a random-coil conformation. When a shearing motion is applied, the random coil opens up and the molecules tend to orient parallel to one another, and in the direction of flow, which is caused by a shearing force. This produces a high viscosity.

1.6   FUNCTIONS OF POLYSACCHARIDES

Polysaccharides are supposed to be the first biopolymers to have been formed on the earth. In some plants, photosynthesized glucose on isomerization produces other hexoses (e.g., fructose, mannose, and galactose), which can also be polymerized into many other plant polysaccharides or oligosaccharides. The galactomannans are one such group of polysaccharides, which are exclusively produced in legume plant seeds.

Polysaccharides have a variety of functions in plants. Thus, cellulose and hemicellulose are essentially the structural components of plants. Other polysaccharides, which accumulate in the seeds, act as a reserve carbon source for a newly sprouted plant from a seed until it starts photosynthesis. Though many seeds produce starch, which is a homopolymer of glucose, as the reserve polysaccharide, many legume plants have galactomannans as their reserve polysaccharide. Galactomannans can hold a large amount of moisture during the water stress period and this appears to be one of their important functions. After isolation from plant seeds, many polysaccharides find applications in human food and in several nonfood industries. These applications, particularly with respect to galactomannans, are covered in the chapters that follow.

As the name suggests, the galactomannan polysaccharides are composed of two types of sugar monomer units, of which mannose is the major component and galactose is the minor component. The number of galactose units in these polysaccharides is always less than that of mannose, and hence, according to the rules of chemical nomenclature of polysaccharides, these are called galactomannans.

FURTHER READINGS

1. Whistler, R. L. and Bemiller, J. N., eds., Industrial Gums, 3rd eds. Academic Press, New York, 1993. (2nd edition, 1973)
2. Aspinall, G. O., Polysaccharides, Oxford University Press, Oxford, UK, 1970.
3. Davidson, R. L., Ed., Handbook of Water Soluble Gums and Resins, McGraw-Hill, New York, 1980.
4. Glickman, M., Gum Technology in Food Industry, Academic Press, New York, 1969.
5. Guthrie, R. D. and Honeyman, J., An Introduction to the Chemistry of Carbohydrates, 3rd ed., Clarendon Press, Oxford, 1968.
6. Arora, S. K., ed., Chemistry and Biochemistry of Legumes, Oxford & IBH Publishing Co., New Delhi, 1982.
7. Wood, W. F., Chemical ecology: Chemical communication in nature, J. Chem. Edu., 60 (1983): 531.
8. Smith, F. and Montgomery, G., Chemistry of Plant Gums and Mucilage, Reinhold Publishing, New York, 1959.
9. Mathur, N. K. and Mathur, V., Industrial galactomannans, Part 1, Guar gum, August 21, 153–162; Part 2, Locust bean and other gums, September 4, 159–162, Chemical Weekly (India), 2001.
10. Lawrence, A. A., Edible Gums and Related Substances, Noyes Data Corporation, Park Ridge, NJ, 1973.

2 Galactomannan Polysaccharides

2.1 INTRODUCTION

Plant seeds have been an ancient source of industrial hydrocolloids or polysaccharide gums. Most of the edible seed grains of the Gramineae family (cereal grasses), for example, wheat, rice, maize, and millet, have starches as their reserve polysaccharides. There are also several legume seeds derived from annual crops (e.g., guar and fenugreek) as well as some full-grown perennial trees and shrubs, (e.g., carob [locust bean] tree and tara shrub) that have endospermic galactomannans as their reserve seed polysaccharides.[1] A reserve seed polysaccharide is that component of a matured seed that does not have other more important biological functions for a plant except to act as a reserve source of carbon for a growing plant embryo before it emerges out of soil and starts doing chlorophyll catalyzed photosynthesis utilizing atmospheric carbon dioxide. Reserve polysaccharides in plant seeds serve as an energy source for man and animals, when used as food.

Anderson, in his extensive survey of galactomannan polysaccharides bearing plants, found that out of 163 species of legume plant seeds, 119 plant seeds contained galactomannans as their endosperm mucilage.[2,3]

Indian flora is particularly rich in legume plants, and legumes form an important component of human food in India. Dr. V. P. Kapoor (retired scientist, National Botanical Research Institute [NBRI] at Lucknow in India) has made an extensive survey of galactomannan-bearing legume plants of the Indian subcontinent.[4,5]

According to R. L. Whistler, who is an authority on industrial gums, "A polysaccharide is classified as a gum, or a hydrocolloid, when it is dispersible in water to form a mucilaginous paste, a colloidal sol or a gel."[6] According to this definition, starch, but not cellulose, falls under the category of gums. Insoluble cellulose, which is not a hydrocolloid, can be converted into its water dispersible derivatives, for example, carboxymethylcellulose (CMC) and methylcellulose, which are gums.

2.2 BIOLOGICAL FUNCTIONS OF GALACTOMANNANS AND THEIR CLASSIFICATION

As mentioned earlier, many legume seeds have galactomannans rather than starches as their reserve polysaccharides. Seed grains, bearing galactomannans generally grow from legume plants in the semiarid regions of the world. Unlike many edible seed grains from the Gramineae family, the seeds of a legume plant (Legumioseae family) are always encased in pods or beans, and these are often embedded in some sort of pulpy material inside the pod. Legume seeds are dicotyledonous, that is, they can be broken into two symmetrical half portions of the endosperm. Legumes in general have more protein (25%–30%) compared to the seeds of grasses (10%–15%).

The biological role of galactomannans in legume seeds is very complex and not very easy to understand. Besides being the reserve polysaccharides, they also provide a proper environment for germination of the seed and guard against water stress during water scarcity periods. They also control and manipulate water absorption by the seeds of different plant species and create a favorable condition and environment for their growth[5] under the prevailing climatic conditions of a geographical region. This property of having a high affinity for water is shared to a varying extent by all the galactomannans, and it has found numerous applications of galactomannans due to which the latter have found many applications in industry and food production.

There are two important groups of galactomannan polysaccharides:

1. Those derived from legume seed-endosperms
2. Those produced by certain microorganisms or bacteria

Bacterial galactomannans (group 2) have more varied and complex chemical structures than the seed galactomannans, but these do not have any commercial applications. In this book I have only dealt with the legume seed galactomannans belonging to group 1, which are commercially produced and find extensive applications.

A number of galactomannans currently being produced on an industrial scale have become commercial commodities.[7] Each one of these galactomannans has some specific properties, which are determined by the molecular weight, the mannose-to-galactose ratio (Man/Gal or M:G), and the mode of placement of single galactose grafts on the linear mannan polymeric chain of the molecule. Hence, one galactomannan may not always be replaceable by another one in a specific application. This is the reason: galactomannans from several different plant sources have to be undertaken for commercial production, in spite of their variable availability and price differences.

Molecular and chemical–structural variations, which resulted in differences in their functional behavior and occurrence, have also resulted in the existence of a broad spectrum of these plant polysaccharides. This has been a windfall for manufacturing and other industries that use these gums. Different galactomannans have provided a large and varied range of applications, which are based on their well-defined structures and functionalities.

In this book an attempt has been made to elaborate, in a comparative way, the manufacture, chemical structure, functional properties, industrial applications, availability, and price structure of commercially produced galactomannans.

2.3 BIOSYNTHETIC ROUTE FOR THE FORMATION OF GALACTOMANNAN POLYSACCHARIDES IN PLANTS[8]

Galactomannan biosynthesis is a photosynthetic process that occurs in many legume plants. This process is in vitro catalyzed by certain enzymes found in plant seed membrane preparations. These membrane preparations were isolated from the developing endosperms of fenugreek and certain other galactomannan-bearing seeds. Galactomannans from plants differ in their mannose-to-galactose ratio, which is, in vivo, 1.1:1 for fenugreek (*Trigonella foenum-graecum*), 2:1 (or more closely to 1.8:1) for guar (*Cymopsis tetragonalobous*), 3.3:1 for senna (*Senna occidentalis*), and 4:1 for locust bean or carob seed (*Ceratonia siliqua*).

The galactomannan molecule is assembled in plants by the action of two seed-membrane-bound enzymes. These enzymes are *mannan synthase* (MS) and *glycosyltransferase* (GMGT). Of these two enzymes, mannan synthase enzyme catalyzes the successive transfer of mannose residues (Man or M) from GDP-Man (a substrate for galactomannan biosynthesis) to an endogenous (presumably the already synthesized galactomannan) acceptor, thus elongating the mannan polymer backbone.

UDP-Gal is the other substrate, which is used in the synthesis of many noncellulosic polysaccharides including galactomannans and certain glycoproteins. The incorporation of galactose (Gal or G) into these macromolecules is catalyzed by *galactosyltransferase* enzyme, which is thought to be localized in the *golgi* apparatus. This galactosyltransferase enzyme from fenugreek seed, which is involved in the synthesis of fenugreek galactomannans, has recently been isolated and cloned.

Thus, the biosynthesis of galactomannans can be represented by the following two basic equations:

M-M--- + GDP-Man (Enzyme, *mannansynthase*) → M-M-M---   (a)

M-M-M--- +UDP-Gal (Enzyme, *galactosyltransferase*) → M-M-M---
$$\qquad\qquad\qquad\qquad\qquad\qquad\qquad\qquad\qquad | $$
$$\qquad\qquad\qquad\qquad\qquad\qquad\qquad\qquad\qquad G \qquad\quad or,$$

M-M-M--- +UDP-Gal (Enzyme, *galactosyltransferase*) → M-M-M--- (b)
$$\qquad\qquad\qquad\qquad\qquad\qquad\qquad\qquad\qquad | $$
$$\qquad\qquad\qquad\qquad\qquad\qquad\qquad\qquad\qquad G$$

Equation (a) represents elongation of the mannan chain, and equation (b) represents addition of the single Gal, graft, at or near the end of mannan chain.

As mentioned earlier, galactosyltransferase enzyme catalyzes the transfer of Gal residues from *UDP-Gal* to form a single galactosyl side chain on the linear mannan chain. Gal residues can be transferred only to an acceptor Man residue at the end, or the Man residue near to the growing (nonreducing) end of an elongating polysaccharide backbone chain.

Further, the transfer of Gal residues conforms to a statistical rule, whereby the probability of obtaining Gal substitution at the acceptor Man residue is determined by the existing states of substitution at the nearest neighbor and the second nearest neighbor Man residues (toward the reducing terminus) of the elongating backbone. This is a second order Markov chain model synthesis. The maximum degree of Gal substitution on the mannan chain, allowed by the deduced Markov probabilities for fenugreek, guar, and senna membranes are very close to those observed for the primary product of galactomannan biosynthesis in vivo. Thus, the biosynthesis of galactomannans requires a specific functional interaction between the two enzymes,

MS and GMGT, within which the transfer specificity of the GMGT is important in determining the statistical distribution of galactosyl residues along the mannan backbone and the mannose-to-galactose ratio.

2.4 GALACTOMANNAN-BEARING PLANTS[1]

The carob tree also known as the locust tree (botanical name, *Cerartonia siliqua*), which is native to the Mediterranean region of Southern Europe and Northern Africa, has been one of the oldest sources of an industrial galactomannan, that is, the locust bean gum or LBG. There are numerous other legume plant seeds bearing galacto-mannan polysaccharides, but from economical considerations only a few of these are currently being used for commercial production of galactomannan gums. From con-sideration of the cost of manufacturing, preferred galactomannan polysaccharides, now under commercial production, are those based on annual plants rather than perennial trees. Putting more land under their cultivation, as and when such need arises, easily increases the cultivation of annual plants. In case of perennial trees, such as those of the carob tree and tara shrub, the fruit bearing occurs only after 8 to 10 years maturing of a tree and then onward the tree continues to bear fruits over a span of nearly 60 years, but the land used in tree plantation cannot be alternated by different crops. Still, due to certain specific functional requirements of a polysac-charide product, many tree seed galactomannans are now commercially produced.

A comprehensive compilation of galactomannan-bearing plants has been made in Dea and Morrison's article titled "Chemistry and Interactions of Seed Galactomannans".[2] The information compiled includes botanical sources, names of the plants and species, percentage of galactomannan in their seeds, mannose-to-ga-lactose ratio, optical rotation data, and water solubility. Among the galactomannan-bearing legumes, soybean (*Glycine max*) is an exception in having galactomannan polysaccharide in its hull or the seed coat, rather than the seed endosperm.

Another important galactomannan-bearing legume is the guar plant (botanical name, *Cyamopsis tetragonolobus*), which is an annual crop. Guar crop has been cul-tivated for centuries, in the Indian subcontinent. Initially guar crop was principally grown to be used as an animal feed and to a much lesser extent unripened guar beans were used as a green vegetable. Guar seed galactomannan was developed as an exi-gency of the Second World War (1939–1946), and it was commercialized only in the early 1950s to be used as a substitute for LBG as a paper pulp additive for the American Paper Industry. During the Second World War, the supply of LBG got severely restricted due to the occupation of most of the LBG-producing Mediterranean region by the Axis countries. Guar gum was discovered at that time as a cheaper alter-native and more dependable source of industrial galactomannan polysaccharide.

Other than LBG and guar gum, there are four other galactomannans that are now commercially produced, raising their total number to six. These are tara gum (from, *Caesalpina spinosa*, which is a perennial shrub of the Peruvian Andes region of South America), fenugreek gum (from *Trigonella foenum-geraecum* L, Hindi name *methi*, which is an annual spice crop now grown world over), *Cassia tora* gum (from a wild, annual herb), and *Sasbania bisipinosa* gum (*daincha* gum in Hindi, which is also a product from a wild annual herb). The last two gums are derived from

TABLE 2.1

Approximate Annual Production Figures of Six Galactomannans

Galactomannan	Production (Metric Tons)
Guar gum	80,000–130,000
Locust bean gum	18,000–20,000
Tara gum	~1,000
Fenugreek gum	8,000–10,000
Cassia tora gum	~5,000
Daincha gum	~5,000

nonagricultured wild, annual herbs. *Cassia fistula* (a perennial tropical tree) is yet another promising galactomannan producing tree, whose gum is, as yet, not commercially produced. Approximate annual world productions of these six gums are shown in (Table 2.1).

Currently LBG, guar, tara, and fenugreek gums are recognized as safe and edible food additives, whereas *Cassia tora* gum is likely to be approved as edible, when it is successfully purified to eliminate possible certain minor toxic constituents present in the commercial product.

2.5 GENERAL STRUCTURAL FEATURES OF LEGUME PODS AND SEEDS[2,9]

Legumes plants produce pods or beans in which varying numbers of seeds are encased. The pod itself is composed of an outer cover in which seeds are enclosed. The seeds in certain pods are surrounded by some sort of pulpy material. Galactomannans are formed in the matured pods of these plants, and the seeds are ready for galactomannan extraction when the matured pods are dry.

The legume seeds are dicotyledonous. They are composed of an outer seed coat or the husk, which is mainly composed of insoluble cellulosic material and pigments. The seed coat encases two, relatively hard and symmetrical endosperm halves, which are mainly composed of layered cell structures composed of galactomannan polysaccharides. The M:G ratio in different layers in the seeds of a particular plant can vary. Between these two seed endosperm halves is sandwiched a protein-rich germ portion or the embryo, which also contains lipids, enzymes, and the genetic material (DNA).

In general, tree seeds are much larger than the seeds from annual crops, producing galactomannans and the hardness of the seed increases as the amount of galactose in a seed galactomannan decreases.

2.6 COMMERCIAL PRODUCTION OF GALACTOMANNANS

Being partially or completely soluble or dispersible in water, the galactomannan polysaccharides are classified as gums or the plant hydrocolloids. Major industrial

TABLE 2.2

Galactomannan Gums Being Commercially Manufactured

Galactomannan	Plant Source	M:G Ratio (Approximate)	Applications
Guar gum	Guar plant (AC)	2:1	F, O
Fenugreek gum	Fenugreek plant (AC)	1:1	F
Locust bean gum	Carob tree (PT)	4:1	F, O
Tara gum	Tara shrub (PT)	3:1	F
Cassia tora gum	*Cassia tora* (wild herb (AC)	5:1	F, (Pet food)
Daincha gum	*Sesbania bisipinosa* (AC)	2:1	O

Note: PT = Perennial tree; AC = Annual crop; F= Food; O = Other.

galactomannans are derived from the seeds of legume plants, which are annual crops, full-grown trees, or shrubs. Bearing of pods or beans as the fruits is a common characteristic of all galactomannan-bearing plants. Legume seeds, being dicotyledonous, are easily split into two halves by coarse grinding. These half portions are commonly referred to as the "split" (*dal* in Hindi). Upon splitting of the seed, the seed coat generally remains attached, sometimes tenaciously to the split.

The legume seed coat, which is called the "husk," is broken and removed to produce the dehusked form of the split. Dehusking of the split is possible, when the seed has been pretreated to increase its moisture content or preheated to make the seed coat brittle. These treatments, followed by some sort of abrasion, shall also remove the husk from the split and this treatment is called "dehusking" of the split.

Thus, commercial production of a galactomannan from any legume seed involves the following steps:

1. Splitting of seed (or making of split or *dal*)
2. Dehusking or removal of the seed coat from the split
3. Separation of the husk and the germ, which are more easily powdered, compared to the intact and hard endosperm, by differential sieving and sifting and separation of purified endosperm
4. Purified endosperm is finally ground to a powder of desired fineness.

Table 2.2 shows important galactomannan gums currently being manufactured, their plant source, Man:Gal ratio, and areas of their application.

2.7   GENERAL CHEMICAL STRUCTURAL FEATURES OF GALACTOMANNAN GUMS[2]

Certain $\beta(1{\to}4)$-linked, linear, and soluble polysaccharides (e.g., galactomannans, glucomannans, and arabinoxylanes) have some common characteristic properties due to their structural similarities. These polysaccharides owe their water solubility or dispersibility to the frequent grafts present on their linear polymeric backbone that prevent, to a large extent, interchain hydrogen bonding, which promotes decreased solubility The legume-seed galactomannans are polysaccharides, having a general structural feature consisting of a linear, polymer backbone of β(1→4)-linked D-mannopyranose units, which are variably substituted by single, α(1→6)-linked D-galactose grafts.

As mentioned earlier, the seed galactomannans from different plants differ widely in their molecular weights, the ratio of the component sugars (M:G), and their functional properties. Thus, the fenugreek seed polysaccharide has nearly all the mannose units of the mannan backbone substituted by galactose grafts, thus making M:G close to 1:1. This is followed by guar gum (2:1), tara gum (3:1), LBG (4:1), and *Cassia tora* gum (5:1), having a decreasing number of galactose grafts. In fact, the different water-solubility fractions of a galactomannan polysaccharide derived from a particular plant source also have variable M:G ratio, and their reported M:G ratio represents the mean average of these ratios in different fractions of the same plant gum.

The following is a typical structural representation of galactomannans:

The viscosity, aqueous solubility and other functional properties, and the applications of a particular galactomannan depend on its molecular weight, besides the M:G ratio, and on the mode of placement of galactose grafts along the mannan polymer backbone. The distribution of galactose grafts can be either uniform, in blocks, or random (see parts a, b, and c, respectively, below). In general, aqueous solubility of a galactomannan increases as the amount of galactose-grafts increase.

There is yet an obvious question: What are the underlying natural reasons for these structural variations in galactomannans from different plant sources? One possible reason could be that these legume plants have evolved and adapted to different soil and climatic conditions of a particular region where they grow, and hence these plants produced galactomannan polysaccharides with different solubility and water-holding capacities.

One common yet important structural feature of the two component sugars (mannose and galactose) in various galactomannans is that both of these sugars have a pair of *cis*-hydroxyl groups in their molecules. These are the C-2, C-3 hydroxyl pair in mannose and C-3, C-4 pair of hydroxyl groups in galactose.

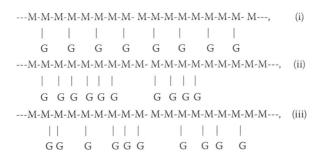

FIGURE 2.1 Possible placement of galactose-grafts in a galactomannan, (i) Regular, (ii) In blocks, and (iii) Random.

The galactomannans, by virtue of the presence of *cis*-hydroxyl groups in their constituent sugars, have a higher affinity for water, compared to cellulose and starch, which are glucose polymers having all *trans*-hydroxyl groups. Some other properties that the galactomannans owe to the presence of the *cis*-hydroxyl group pair include their interactions with borax at alkaline pH to form gel and formation of complexes with certain transition metal ions. The *cis*-hydroxyl pair also forms cyclic carbonate ester with phosgene and cyclic acetals with aldehydes (see Chapter 4, Section 4.11).

There are many other functional properties of galactomannans where these polysaccharides differ from those of glucose polymers, and these can also be attributed to the presence of *cis*-hydroxyl groups of their numerous components sugar monomers. Surprisingly no literature description of galactomannan polysaccharides, other than the present one, has tried to correlate properties of galactomannans to the presence of *cis*-hydroxyl pairs in these polysaccharides. These specific properties of galactomannans are discussed in detail in Chapter 3.

2.8   GLUCOMANNANS

Besides galactomannans, glucomannans are yet another group of polysaccharide hydrocolloids derived from plants. Glucomannans are known to occur in aloe vera leaves and in the konjak tuber. Mannose is the common sugar component in galactomannans and glucomannans, and they share a common property of having a very high affinity for water. Structural similarity between these two groups of polysaccharides lies in their linear, $\beta(1\rightarrow4)$-linked pyranose sugar blocks of mannose interposed by those of glucose, and short, single sugar (glucose) and acetyl grafts.

Glucomannans can have some of the hexose units of their polymer backbone, which are di-acetylated or even tri-acetylated, making these units less hydrophilic. Because of some structural and functional similarities, glucomannans are also briefly described in this book.

2.9   COMMERCIAL NEED OF GALACTOMANNANS HAVING DIFFERENT FUNCTIONAL PROPERTIES

As mentioned earlier, guar gum was initially introduced as a paper pulp additive for the U.S. paper industry to substitute for LBG. Soon afterward it also found many applications in other fields, for example as a thickener for the mud for oil-well drilling, food additive, slurry explosives, textile printing paste thickener, and forest fire control agent. Guar gum consumption got a further boost when its anionic, cationic, and hydrophobic derivatives were synthesized, which immediately found numerous industrial applications.[10,11]

Upon cooling, even concentrated solutions of galactomannans do not form true gels. But when a galactomannan, such as locust bean gum, is mixed with certain gelling polysaccharides, such as agar or carageenan (both seaweed polysaccharides), the mechanical strength of the binary gels, thus formed is improved. This is due to the synergistic interaction of galactomannans having a lower percentage (<25%) of galactose sugar units in their molecules. With guar polysaccharide, having more galactose monomers (>33%) synergy only results in an increase in viscosity of several soluble, linear polysaccharides.

Being a cheaper and abundant galactomannan compared to LBG, guar gum has replaced LBG in most of its nonfood (technical) applications. Yet, there has also been an increasing demand in the food industry for LBG and similar gums that are lower in their galactose content. Galactomannans, having a lower galactose content (M:G = 3:1 or less), have many applications as additives for food. This is because of their lower water solubility and a mixed sol-gel rheology. This will be discussed in Chapter 3. Since the production of LBG has been stagnant, many of its substitutes are being explored, and *Cassia tora* gum is likely to emerge as one such product.

2.10   FUTURE PROSPECTS[12,13,14]

Many more seed galactomannans bearing plants, such as *Cassia fistula*, has been identified. Future prospects for more plant seeds being commercially exploited as sources of industrial galactomannan gums will depend on the functional properties of their gum, their availability, and price. Today guar gum is the most abundantly produced galactomannan, which is being used in food as well as nonfood applications. Variable annual production of guar seed, from its rain-fed crop, as it was some two decades back, and consequently a large fluctuation in its market price, has now been very much reduced. This has been possible due to its irrigated crop production in some regions. The production of fenugreek seed is also posed to increase, and use of its gum in low calorie, diet food is now increasing. Development work now being carried on *Cassia tora* galactomannan to process it into food-grade quality is certainly going to produce a good substitute of LBG. Gums from trees having limited production are less likely to be used in nonfood applications.

In India, particularly in Jodhpur (Rajasthan state), some larger industrial groups of companies, which were initially dealing only in guar gum production, have now turned into multigalactomannan or even multihydrocolloid companies. These companies have very good research and development staff and quality control facilities. These industries are no more dependent on foreign know-how and are marching ahead to develop into larger hydrocolloid companies.

REFERENCES

1. Reid, J. S. G., Galactomannans from legume seeds endosperm, Bot. Re., 11 (1985): 125–155.
2. Dea, I. C. M. and Morrison, A., Chemistry and interactions of seed galactomannans, Adv. Chem. Biochem. Carbohyd., 31 (1975): 241–312.
3. Anderson, E., Ind. Eng. Chem., 41: (1949): 2887.
4. Duke, A., ed., Handbook of Legumes of World Economic Importance, Plenum Press, London, 1981.
5. Kapoor, V. P., Sent, A. K., and Farooqi, M.Z. Indian J. Chem., 28B (1989): 281.
6. Whistler, R.L. and BeMilter, J.N., Eds. Industrial Gums, 3rd ed., Academic Press, New York, 1993, 9.

7. Cui, S. W., Polysaccharide Gums from Agricultural Products, Processing, Structure and Functionality, Technomic Publishing, Lancaster, PA, 2002.

8. Whistler, R. L. and Hymowitz, T., Guar: Agronomy, Production, Industrial Use and Nutrition, Purdue University Press, West Lafayette, IN, 1979.

9. Stephen, A. M., Phillips, G. O., and Williams, P. A., Food Polysaccharides and Their Applications, 2nd ed., CRC Press, Boca Raton, FL, 2006.

10. Paroda, R. S. and Arora, S. K., Guar—Its Improvement and Management, The Indian Society of Forage Research, Hissar, India, 1978.

11. Smith, F. and Montgomery, G., Chemistry of Plant Gums and Mucilages, Reinhold Publishing, New York, 1959.

12. Glickman, M., Gum Technology in Food Industry, Academic Press, New York, 1969.

13. Mathur, N. K., Narang, C. K., Sharma, I. K., and Mehra, A., Polysaccharides and functional polysaccharides: Preparation and applications. React. Polym., 6 (1967): 73.

14. Mathur, N. K. and Mathur, V., Industrial Galactomannans, Part 1, Guar Gum, August 21, 153–162; Part 2, Locust Bean and Other Gums, September 4, 159–162, Chemical Weekly (India), 2001.

3 Hydrocolloids or Gums

3.1 DEFINING A COLLOIDAL SYSTEM, POLYMERIC AND NONPOLYMERIC COLLOIDS

There are many substances consisting of particles that are substantially larger than atoms and ordinary molecules, but still much too small, to be visible to the unaided eye or under an optical microscope. These are referred to as *colloidal systems* or simply as colloids. A simple *colloidal solution* or a *sol system* is generally defined as a system of such small particles of a solid dispersed in a liquid phase. More generally, a colloidal system may exist as dispersions of a substance in one phase (called the *dispersed phase*, which can be a solid, liquid, or a gas) into a different phase, called the *dispersion phase*. As an example, mist consists of tiny liquid water droplets dispersed in air (gas phase), whereas smoke is a colloid system in which solid particles (carbon, soot) are dispersed in air. Dispersions of plant gums and rubber latex are colloid systems having molecules of a plant polysaccharide or a polymeric rubber hydrocarbon, both of which are solids, dispersed in a liquid water phase. All these are biphasic systems, where the dispersed particles are much smaller in size than the coarse particles. The dispersed particles are either polymeric macromolecules, or larger aggregates of smaller atoms or molecules, which are not visible to the naked eye or even with the help of an optical microscope.[1]

In a nonpolymeric colloidal system, the suspended particles are aggregates of small molecules or atoms, the size of the aggregate approaching colloidal dimension, which is considered to be in the range of 5 to 200 millimicrons. These aggregates form a stable dispersion in a liquid, which can remain in suspension for a considerable length of time (i.e., several days) without settling under gravitational force. Many colloidal systems consist of charged particles and mutual charge repulsion prevents their aggregation into still larger (noncolloidal) ones. One essential characteristic of colloid particles is their large surface area in ratio to their volume. Colloidal suspensions behave as a heterogeneous, two-phase system: the dispersed phase and the dispersion phase. The colloid particles are large enough to scatter the light beam falling on them.[2]

3.2 POLYMERIC COLLOIDAL SYSTEMS AND THE SIZE OF COLLOIDAL PARTICLES

The lower limit of optical microscope visibility for particles is ~200 millimicrons. This is generally considered as the upper limit of the size of colloidal particles, while the lower limit of the colloid particle size is ~5 millimicrons. This lower sizelimit is also the approximate range of the size of many complex and high molecular weight polymers, and natural macromolecules (e.g., the molecules of globular proteins, starch, and many polysaccharide gums, rubber hydrocarbons, and synthetic polymers). A sol system of this kind, that is, a solution or a dispersion of a macromolecular substance, is a colloidal system, where the colloidal particles generally exist as separate molecules of a polymeric material. These polymeric molecules are of the size of colloidal particles. Typical examples are milk proteins and solutions of plant gums, and synthetic polymers, polyelectrolytes, and blood proteins. The size of polymeric molecular colloids is much smaller, ~5 millimicrons, compared to colloids, which are molecular aggregates.

3.3 REVERSIBLE AND IRREVERSIBLE COLLOIDAL SYSTEMS: HYDROCOLLOIDS

Colloid systems are generally classified into two groups: reversible and irreversible colloids. In a reversible colloidal system, the products of a physical change (e.g., dissolving of a water-soluble polysaccharide) or a chemical change resulting in a colloid sol formation may be reversed in such a way as to reproduce back the original components. In a system of this kind, the colloidal materials are generally high molecular weight polymers or the macromolecules, where a single polymeric molecule is of a colloidal dimension.

An irreversible colloid system is a nonpolymeric type. It consists of relatively stable aggregates of nonpolymeric substances dispersed in a liquid phase. In case of an irreversible colloid system when the dispersed colloid phase (which is otherwise stable) is removed from the dispersion phase, then its original components cannot be reversed back to the colloidal state simply by putting it in the dispersion phase. Examples of irreversible systems include sols of clay suspensions at high dilution; pastes, which are rather concentrated suspensions; emulsions; foams; and certain varieties of gels. The size of the particles of these colloids is greatly dependent on the method of preparation employed.

Macromolecular substances of colloidal dimension, which readily disperse into an aqueous phase, are referred to as *hydrophilic polymers* or simply *hydrocolloids,* and these form a reversible colloid system. The term *gum* is used for many commercial hydrocolloids, mostly derived from polysaccharides.[3] The terms gum and hydrocolloid are often used interchangeably. Since the hydrocolloids are reversible colloids, it means that after their crystallization from water, they can again be dispersed into colloidal form by putting them into water. Most of the hydrocolloids are hydrophilic polymers or gums of vegetable, animal, microbial, or synthetic origin. They contain large numbers of hydroxyl, amino, carboxylic, and many other polar or ionic and hydrophilic groups, by which they can interact with water.

Hydrocolloids find many applications in food production and several nonfood industries. Some hydrocolloids are naturally present in food, while many others are added to food to control the functional behavior of foodstuffs.[4,5] Most important of these functional properties are viscosity control, gelling, and stabilization of solid or foam suspensions and emulsions.

3.4 CHEMICAL STRUCTURE OF POLYSACCHARIDE HYDROCOLLOIDS[6,7]

Polysaccharides are classified and named according to their sugar components; their sequences and linkages between them as well as the anomeric configuration of linkages; the ring size of their constituent sugars, that is, furanose or pyranose; the configuration of component sugars (D or L); and any other substituents present in them. Certain structural characteristics such as chain conformation and intermolecular associations shall also influence the physicochemical properties of polysaccharides.

Linear and grafted linear polysaccharides-based hydrocolloids are often represented by an idealized structure or a repeat unit (i.e., amylose), which is the linear fraction of starch, and guar gum are represented by the following idealized structures (see also Figure 3.1).

$$- (\text{-Glc-}\alpha\text{-1-4-Glc-})_n - \qquad \text{(a)}$$
$$-[(\text{-Man-}\beta\text{-1-4-Man-})]_n - \qquad \text{(b)}$$
$$|\alpha\text{-1-6}$$
$$\text{Gal} \qquad \text{(c)}$$

Part (a) is a fragment of an amylose molecule, (b) is a fragment of a guar gum molecule, and (c) is a fragment of a guar gum molecule presented in the pyranose structure. In the pyranose structure representations of a fragment of guar gum (c), the oxygen atoms, which are shown in gray, are the sites of hydrogen bonding. These can interact and bind water molecules or their clusters. Due to the flexibility of the linear chain of these molecules, in a solution these linear polysaccharide molecules generally exist as a random coil, that is, a randomly folded structure. On applying a shearing force, a random coil molecule can open up and acquire a linear or rodlike conformation.

Highly branched gums (e.g., gum acacia and an amylopectin fraction of starch molecules) are spheroid or globular in shape. These have very little sequential

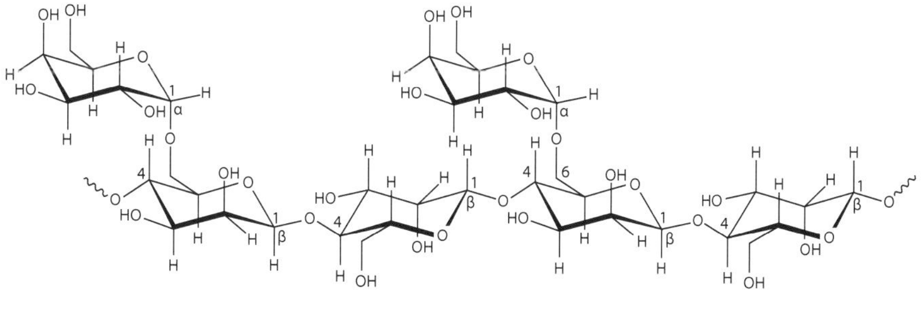

FIGURE 3.1   An idealized, structural representation of galactomannan (guar-gum) having Mannose: Galactose ratio of 2:1.

placement of sugar monomers along their short braches. These cannot be represented by idealized, repeat unit structures similar to the linear polysaccharides. Overall, the polysaccharides are natural products whose structures are determined by stochastic, biosynthetic reaction, that is, random enzymatic action. These structures are not laid down exactly by any specific genetic code, as in the case of functional proteins. Hence, no two molecules of a particular polysaccharide may be exactly identical in their molecular weight, structure, and conformation.

3.5 POLYDISPERSIBILITY AND CONFORMATION OF HYDROCOLLOIDS

As stated earlier, the industrial polysaccharides (gums) are made up of mixtures of molecules with different molecular weights, and no two molecules in such a mixture are likely to be conformationally and structurally identical. Such polymers are referred to as *polydispersed materials*. Many of the linear hydrocolloids, particularly the charged ones or the polyelectrolyte, retain their extended structures in solution; whereas uncharged, linear molecules exist as random coil structures. Depending upon the structure of the monomeric sugar units in a carbohydrate polymer, it can also acquire a spiral or a helical conformation (e.g., carrageen, amylose, and DNA). This may also give rise to a mixed entanglement or a phase-separated entanglement of linear molecules. At a concentration higher than a "critical concentration," these polymers may produce gels. Phase separations may be entropy driven, as they may allow greater freedom of movement due to the similarity in the molecular shapes. Sometimes *junction zones* are produced between linear polymer molecules, resulting in gel formation.

3.6 FUNCTIONAL PROPERTIES OF HYDROCOLLOIDS AND THEIR STRUCTURE–FUNCTION CORRELATION

The most stable arrangement of atoms in a polysaccharide will be that which satisfies to a maximum extent both the intra- and intermolecular hydrogen bonding forces. Regularly ordered polysaccharides are capable of assuming only a limited number of conformations due to severe steric restrictions on the freedom of rotation of sugar units around the intersugar unit, glycosidic bonds.

Thickening and gelling, in the aqueous phase, are two important functional properties of hydrocolloids. Mixtures of hydrocolloids show a complexity of these nonadditive properties, which have been difficult to interpret. It is only recently that such functional properties (e.g., synergy in viscosity enhancement of a mixture of polysaccharides and binary gelling of polysaccharides) have been interpreted as a science rather than an experience-based art.

Hydrocolloids can be dispersed in liquid water, which by itself has a very complex structure. Hence, there is an enormous potential in combining the structure–function knowledge of polysaccharides with that of the structuring of water. Particular parameters of each application of a hydrocolloid in any food or nonfood industry must be examined carefully, considering the effects that are required of it (e.g., to improve the texture, viscosity, bite and mouthfeel, water content, stability, stickiness, cohesiveness, resilience, springiness, extensibility, processing time, and process tolerance) in each of the cases. In a mixture of polysaccharides, synergistic enhancement of viscosity or binary gel formation can be observed. Due regard is also given to the type, source, cost, purity grade, and structural heterogeneity of a polysaccharide hydrocolloid.

Particle size and particle size distribution of a commercial hydrocolloid sample are important parameters, which determine rate of its hydration, dissolution, and its emulsification ability. Hence, particle size distribution in a gum powder can have an effect on its functional properties.

Negatively charged hydrocolloids change their structural characteristics with the type of counterion present and its concentration in a sol. The pH and ionic strength also effect their functioning, for example, at high acidity the anionic charge on a hydrocolloid can disappear and its molecules become less extended or randomly coiled in shape. Physical characteristics of hydrocolloids are controlled by thermodynamics as well as by a kinetics consideration, and hence its processing history and environment in which it is present, determine its required concentration in a specific use. In particular these factors may change with time in a monotonic or oscillatory manner.

3.7 STRUCTURE–SOLUBILITY RELATIONSHIP IN HYDROCOLLOIDS

As the name suggests, hydrocolloids have a large affinity for liquid water. Temperature and pressure of an aqueous sol system also influence the extent of hydrogen bonding interactions in it. Charged or polyanionic hydrocolloids are more soluble in water than nonionic ones. Hydration and solubility of hydrocolloids depends on several factors. Thus xanthan, carboxymethylcellulose (CMC), guar, and fenugreek gums are soluble in cold water, whereas carrageenan and locust bean gum (LBG) dissolve only in hot water. On dissolving, water clusters are reversibly held by hydrogen bonding to the hydrocolloid molecule. This means that structuring of water takes place within inter- or intramolecular spaces of the polymer molecules. A hydrocolloid can have several different conformations in a solution due to its rotational motion and translocation of its polymeric chains.

CMC is a typical hydrocolloid that can be synthesized from a nonhydrocolloid, that is, a water-insoluble polysaccharide, cellulose, by carboxymethylation reaction. Introduction of anionic carboxymethyl (CM, or $-CH_2COO^-$) side chains in a linear and insoluble and nonionic cellulose polymer makes it water soluble. The average number of carboxymethyl groups per monosaccharide unit introduced in a polysaccharide molecule can be controlled during its synthesis, and it is called the *degree of substitution* (DS). In case of galactomannans, having on an average three hydroxyl groups per monosaccharide unit, the maximum DS can be three. At higher DS, unfolding of linear molecular chains increases and these are held apart due to charge repulsion. The anionically charged side chains also prevent formation of interchain hydrogen bonds and increases solubility. Thus, anionically charged CM–LBG hydrates and disperses much better compared to the original

LBG, which is a nonionic polysaccharide. Uniformity of distribution of these CM groups at low DS along the polymeric chain also determines the solubility of a hydrocolloid.

Nonionic galactomannan molecules have several single galactose grafts on the linear mannan backbone. In LBG the distribution of grafts is still more uneven compared to that in guar gum. This reduces the hydration rate and solubility of LBG. Thus, LBG does not dissolve in cold water; it only swells. In contrast to this, guar gum is cold-water soluble. LBG with uneven distribution of side groups interact synergistically with xanthan gum, which is a nongelling polysaccharide, to form a gel. In contrast to this, guar gum does not form a binary gel with xanthan, and only an increase in viscosity takes place. This means that good cold-water solubility is not always a desired property of a food hydrocolloid.

Hydrolytic enzymes can disrupt and depolymerize the covalent structure of a hydrocolloid, which in turn affects their performance. Gums are also hydrolyzed thermocatalytically as well as by acids and alkalies, which results in their reduced viscosity.

3.8   INTERACTION OF HYDROCOLLOIDS AND WATER[6,8]

A linear hydrocolloid molecule in water can be visualized as a flexible polymer string and surrounded by a cylinder of water. This water is associated by hydrogen bonding to the thin, linear polymer molecule and, to an extent, it will move along with it when shearing (stirring or flowing of the sol) takes place. When anionic groups are present on the polymer backbone, they repel one another and cause chain elongation. Thus, elongated polymer exhibits enhanced viscosity. The basic activity of a hydrocolloid is attributed to its ability to bind clusters of water and to form a network of its chains. A hydrocolloid's interaction with water also slows down its diffusion in a sol and stabilizes water structure. Nonionic hydrocolloids are less soluble in water, whereas the hydration kinetics of a polyelectrolyte depends on many more factors.

Water molecules in a sol may be held specifically through direct hydrogen bonding or by the structuring of water within inter- and intramolecular voids of the dispersed polysaccharide. As mentioned earlier, hydrogen-bonding interaction between a polysaccharide and water is dependent on the temperature and pressure in the same way as water cluster formation in pure water. There is a reversible balance between the entropy losses and enthalpy gains, but the process may be kinetically limited and an optimum hydration of a polysaccharide network may never be reached.

Hydrocolloids exhibit a range of variable conformations in an aqueous sol when the links between the sugar monomer units of polysaccharide chains can rotate relatively freely. Large and conformationally stiff polysaccharides present essentially static surfaces, which can cause extensive structuring of the water surrounding it.

Water binding affects the texture and processing characteristics of a hydrocolloid. This in turn can cause, or prevent, syneresis (water separation) from a gel or a sol. Water binding of gums can bring substantial economic benefits to the polysaccharide user in many of its applications. Specifically the hydrocolloids can provide water for increasing the flexibility or plasticizing of food components. Hydrocolloids can also minimize or completely prevent ice crystal formation and thus have favorable influence on the texture of frozen foods (e.g., ice creams). Some hydrocolloids, such as a mixture of LBG and xanthan gum, form strong gels on the freeze–thaw cycle due to certain kinetically irreversible changes. As a consequence of this, any forced association, such as water activity, is reduced on freezing.

The chain length of a linear hydrocolloid influences its viscosity and hydration rate. Longer molecules tend to produce higher viscosities but may take a longer time to fully hydrate. A highly branched polysaccharide molecule takes up less space of gyration in a sol than a linear one with nearly the same molecular weight. A highly branched chain polysaccharide molecule (e.g., acacia gum), therefore, provides a much lower viscosity.

3.9 BINDING OF MOLECULAR CLUSTERS OF WATER BY HYDROCOLLOIDS[4,5]

Polysaccharides are more hydrophobic when they have a large number of intramolecular hydrogen bonds. The increased hydrophobic nature of a polysaccharide results in reduced interaction with water. Carbohydrates contain hydroxyl groups, which preferentially interact with two water molecules each, provided they are not interacting with other hydroxyl groups of the same molecule. Interaction with hydroxyl groups on the same or neighboring sugar residues will necessarily reduce the polysaccharide's hydration status. β-Linkage to the 3- and 4-positions in mannose or glucose homopolymers allows strong inflexible, interresidue hydrogen bonding, which reduces overall polymer hydration. It gives rise to rigid inflexible structural polysaccharides. In contrast, α-linkages to the 2-, 3-, and 4-positions in mannose or glucose polymers give rise to greater aqueous hydration and more flexible linkages.

Water binding by hydrocolloids arises due to hydrogen bonding of polar or ionic groups in a polymer to the solvent water molecules. This is a significant property of hydrocolloids, which can result in emulsion stabilization, prevention of ice crystal, and organoleptic (taste and mouthfeel) properties of food. Industrial applications of hydrocolloids include viscosity control, adhesion, suspension, flocculation, foam stabilization, and film formation.

Examples of common hydrocolloids used in food are agar, carrageenan, and alginate (from seaweeds), arabinoxylanes (from wood hemicellulose), CMC (modified cellulose gum), curdlan, xanthan and gellan (from microbial origin), gelatin (protein from animal hides and skin), starches and cell-wall beta-glucans (from food grains such as wheat), locust bean gum, guar gum, and other galactomannan polysaccharides (from legume plant seeds), gum arabic (tree exudates), and pectin (gelling constituent of fruit skins). All these hydrocolloids, except gelatin (a protein), are polysaccharides. Being polyhydroxy or polycarboxylic compounds, polysaccharides are strongly hydrated in water.

Hydrocolloids–water interaction results in reducing the diffusion of water, which ultimately results in the stabilization of water structure. Most linear hydrocolloids, at low concentrations, influence large volumes of water within the radius of their gyration. As the concentration of a gum in a sol is increased, different scenarios can be seen. Linear molecules of a polysaccharide may fold up into a globular structure or random coil, resulting in the decrease of gyration volume and viscosity. This is an entropy-driven process, which can still allow the retention of their rotational freedom. Further increase in concentration of a polysaccharide in the dispersed phase may cause phase separation and gelling due to different influences on the water-cluster structuring and activity of water as a solvent.

The presence of a polysaccharide hydrocolloid in a sol increases the viscosity of the medium (water), which affects its flow behavior as a solvent, although water is present many folds the weight of the solute polysaccharide. Hence, most of the hydrocolloids are used to increase the viscosity and to modify the rheology of an aqueous system. Increase in viscosity, in turn stabilizes foodstuffs or other nonfood systems by preventing or minimizing settling of suspended solids, phase separation, foam collapse, and crystallization of water on lowering the temperature. The subject of a the rheology of polymer sol is discussed in greater detail in Chapter 4.

The viscosity of a hydrocolloid sol changes with its concentration, temperature, and shear strain rate of its sol in a very complex manner. Viscosity of a sol also depends on the nature of a hydrocolloid and other materials present in it. Mixtures of hydrocolloids frequently act synergistically to increase the viscosity, which is much beyond an expected value. In some other cases, polysaccharides may act antagonistically to reduce the anticipated viscosity. Due to the recent advances in the science of rheological studies, many of these changes can now be fitted into semiquantitative equations.

For a particular application, the choice of a hydrocolloid depends on the functional behavior required in a finished product. Thus, the knowledge of a required specific rheological characteristic of a system helps in determining the choice of a hydrocolloid, which can provide the necessary viscosity to a sol and elasticity or hardness to a gel.

3.10   GELS AS COLLOIDAL SYSTEMS AND GEL-FORMING HYDROCOLLOIDS[9]

Most hydrocolloids only thicken a sol, while a few of them can cross-link the dispersed polymeric molecules into aggregates, by multiple secondary bonds (hydrogen bonding) and dipolar interactions, to form junction zones resulting in the formation of a three-dimensional network called a gel. A gel is a viscoelastic structure. Some food gels are thermoreversible, that is, they will liquefy into a sol on heating. In gel formation, intra- and intermolecular hydrogen bonding is favored over hydrogen bonding to the solvent water, to an extent, which is sufficient to overcome the entropy factor (flow of water).

Gels too form a biphasic colloid system. A gel is a coherent mass appearing to be a semisolid and consists of a liquid phase in which solid particles of colloidal dimensions are either dispersed or arranged in a fine network throughout the mass. Some food gels are notably elastic and jellylike (e.g., gelatin and fruit jellies). Some other nonfood gels can be solidlike and rigid to feel (e.g., the silica gel), which is used as a dehumidifier. In case of an elastic food gel, the liquid media in the gel system becomes viscous enough to behave more like a semisolid. Contraction of such a gel can cause separation out of liquid from the gel, which is termed *syneresis* of a gel.

Galactomannans do not form true gels, yet partially soluble galactomannans can have a semigel-like appearance in a more concentrated solution. Gel-forming hydrocolloids (e.g., agar, carageenan, and xanthan) form more economical binary gels in the presence of certain galactomannans.

Gel formation by hydrocolloids has several important applications in the food industry. Hydrocolloid gel can be considered as a glassy state of material, whereby the conformational changes of polymeric molecules are severely inhibited. Water held by the hydrocolloid molecules in a gel acts as a plasticizer, which allows molecular motion but reduces the glass-transition temperature by inhibiting intermolecular hydrogen bonding.

Hydrocolloids exhibit a delicate balance between a hydrophobic and hydrophilic character. Hydrophobic interactions bring about clustering of identical groups, located in different parts of the same molecule or on different molecules. Linear hydrocolloid molecules have the tendency to entangle at higher concentrations. Similar molecules may be able to wrap around one another, forming helical junction zones. When hydrogen bonding is retained during a helix formation, conformational heterogeneity is reduced, minimizing hydrophobic surface contact with water. Water is thus released energetically for favorable use elsewhere in the system. Under such conditions, a minimum number of interpolymeric links may need to be formed. Helical junction zones generally require a complete helix to overcome the entropy effect and form a stable link. Where junction zones grow slowly with time, these interactions eliminate water and syneresis (creeping out of fluid from a gel) may occur. In the case of gelling systems involving low-methoxyl pectin or alginate, metal ions (e.g., those of calcium) are required to form a gel.

Gel strength is defined as the force to break a gel. Several commercially available instruments can measure it. Thixotropic gums form a weak gel, which is broken when the applied shear equals to the yield point. If the junction zones expand with time, the gel structure contracts and therefore squeezes out the bonded water, resulting in syneresis. Such rheology is related to the structure of the molecules and has useful applications designing food products.

Hydrocolloid gels are considered to be fluids containing a polymer network and showing solidlike behavior with characteristic strength, which depends on their concentration. Hardness or brittleness of a gel depends upon the structure of the hydrocolloids present in it, which controls textural properties of many foods. Gels display elastic as well as viscous behavior. Elasticity occurs when the entangled polymers are unable to disentangle in time to allow a flow. Mixtures of hydrocolloids may act synergistically, associating to precipitate as a gel or form incompatible biphasic systems. Such phase confinement affects the viscosity as well as the elasticity of the system.

Hydrocolloids are versatile, when they are used for many other purposes including the production of pseudoplasticity, that is, fluidity under shear.

Heating and vigorous stirring of a gum in water generally makes hydrocolloid sol to ease its mixing and dissolving. Upon cooling, the resulting sol generally thickens, and gelling takes place in some cases. In case of thermoreversible gel, liquefaction results on heating and gelling on cooling. Typical examples of such cold-setting gels are those formed by seaweed gums (e.g., agar, gellan, and carrageenan), which are further reinforced by galactomannans.

3.11   SYNERGY AMONG HYDROCOLLOIDS[10]

It was mentioned earlier that a nongelling hydrocolloid can form a binary gel in the presence of another nongelling hydrocolloid. A mixture of two gums in a solution can also result in a synergistic effect on viscosity. For example, when a solution of a 1% CMC (viscosity ~3,000 cps) and an equal amount of 1% guar gum solution (viscosity ~3,000 cps) is considered, we might expect the viscosity of this mixture also to be ~3,000 cps. Actually the viscosity increases to a range of 5,000 to 6,000 cps. This is due to random collisions resulting in association of the hydrated linear polymer molecules in the mixture. In some cases this can also result in binary gel formation. In case of a gelling gum (e.g., carrageenan) upon mixing it with a nongelling LBG, the gel strength is improved. Increased gel strength is also possible when the thickening gum has an additional water-organizing capacity. Linear and extended chain polymers in a solution tend to entangle with each other at a higher concentration, forming a helix and reducing conformational heterogeneity and minimizing hydrophobic surface contact with water.

Synergy provides an economic benefit to the polysaccharide end user. This can also help to change the texture of a product produced by a gum. An anionic and a cationic gum are mutually incompatible when present in a solution in molar ratio, and precipitation can occur. At a very low molar ratio, a cationic gum is sometimes deliberately added to anionic cosmetics, for example, in a shampoo, in order to increase its viscosity and to improve its moisturizing action.

3.12   EMULSIFICATION BY HYDROCOLLOIDS

Most of the polysaccharide-based hydrocolloids stabilize emulsions due to an increased viscosity of aqueous sol and slowing the thermodynamically favored breakdown of the emulsions, which are colloidal systems containing two immiscible liquids. Some hydrocolloids by themselves also act as emulsifiers and their emulsification abilities are reported to be due to the contamination of hydrophobic protein moieties present in them. Thus gum arabic, which is composed of a hydrophilic polysaccharide, has some hydrophobic protein parts bound to it. Electrostatic interaction between ionic hydrocolloids and proteins may give rise to a marked emulsification ability with considerable stability at appropriate pH and ionic strength. A typical example is denaturation and coagulation of the milk protein, which is likely to lead to improved emulsification ability and stability. Among the galactomannans, highly purified fenugreek gum (free from protein impurities) has good emulsifying properties, the mechanism of which will be dealt with in Chapter 8.

3.13   APPLICATIONS OF GALACTOMANNAN HYDROCOLLOIDS IN FOOD AND NONFOOD INDUSTRIES[11]

Hydrocolloids have important roles in many industries, more so in food, which is basically an aqueous system. In an aqueous system, hydrocolloids change a product's texture and viscosity, along with a wide range of other special effects, for example, they can cause gelling. The food industry makes ample use of many multifactorial and multifunctional hydrocolloids. The methods of preparation of food, its thermal processing, and foodstuff environment (e.g., the presence of salt, fat, acidity, or the pH and temperature) affect functioning of a polysaccharide used. In food items, hydrocolloid gums are used to influence the texture, bite, mouthfeel, and water binding to prevent its crystallization in ice cream and confectionery products. Hydrocolloids control the distribution of solid particles in an aqueous sol, such that a large variety of textures and mouthfeel can be produced. By judicious use of polysaccharides, a firm, elastic, or a hard gel can be produced in confectionery products.

As mentioned earlier, some hydrocolloids form a gel, whereas others act only as thickeners. Fully hydrated polysaccharide molecules exhibit little interaction with other molecules present in a system. There is lot of rheological differences between different gums, which make them suitable for particular food or nonfood applications. With those gums that cause only thickening, distinctive flow properties in a sol are still maintained in food systems. When a gum used is thixotropic in nature, it produces a less slimy texture and such gums are preferred as food additives. Temperature effects on the viscosity and gelling in a food system are generally reversible. The gum being used in food applications should be able to withstand the processing of food and should not decompose during the cooking process. Thickeners are used in sauces, dressings, beverages, soups, and other food applications that require thickening, and to physically bind water in a fat-reduced product. A thermally reversible gel would be used in water-based or milk-based gel desserts as well as in canned meats and some confectionery products. Thermoirreversible gels are used in jellies, jams, gelled confectionery products, and restructured foods. Both thickeners and gelling gums stabilize products by structuring water and stabilizing of emulsions.

In the petroleum industry, galactomannans are used in drilling mud, for suspending clay, and for enhanced oil recovery. Water-based paints, construction materials (e.g., cement formulations), textile printing pastes, slurry explosives, and firefighting are some nonfood industries that use galactomannans and other hydrocolloids. More detailed and specific applications of each galactomannans are discussed in other chapters.

REFERENCES

1. Alexander, A., Colloid Chemistry, Reinhold Publishing, New York, 1944.
2. Glasstone, S. and Lewis, D., Elements of Physical Chemistry, 2nd ed., Macmillan & Co. Ltd. London, 1960.
3. Garti, N., Food hydrocolloids, 8(2) (1994): 155.
4. Graham, H. P., Ed., Food Colloids, Avi Publisher, Westport, CT, 1977.
5. Eisenberg, D. and Kauzmann, W., The Structure and Properties of Water, Oxford University Press, New York, 1969.
6. Whistler, R. L. and BeMiller, J. N., Eds., Industrial Gums, 3rd ed., Academic Press, New York, 1993.
7. Aspinall, G. O., Polysaccharides, Oxford University Press, Oxford, 1970.

8. Jeffrey, G. A. and Saenger, W., Hydrogen Bonding in Biological Structures, Springer Verlag, Berlin, 1991.
9. John, F., Viscoelastic Properties of Polymers, Wiley, New York, 1980.
10. Dea, I. C. M., and Morrison, A., Chemistry and interactions of seed galactomannans, Adv. Chem. Biochem. Carbohyd., 31 (1975): 241–312.
11. Glickman, M., Gum Technology in Food Industry, Academic Press, New York, 1969.

4 Interactions of Galactomannans

4.1 INTRODUCTION

Interchain and intrachain hydrogen bonding interactions of cellulose (a homopolymer of glucose), which have resulted in its insoluble and fibrous nature, are well documented in the literature. Structure–interaction correlations for cellulose have also been well interpreted to explain its crystallinity. Hydrogen bonding interactions related to certain soluble polysaccharides (i.e., galactomannans and glucomannans) have also been referred in literature[1,2] related to these carbohydrate polymers.

It is the presence of *cis*-hydroxyl pairs in the component sugars of galactomannan and glucomannan polysaccharides that makes a major difference in their interactions when compared to those in cellulose. Cellulose and starch, which are two glucose homopolymers, have all *trans*-hydroxyl pairs in their component sugar, glucose. There is also much difference in the functional properties and interactions of soluble cellulose derivatives (e.g., carboxymethylcellulose [CMC]) on one hand and galactomannans on the other hand. Both these groups of polysaccharides have a backbone composed of $\beta(1\rightarrow4)$-linked pyranose sugars.

Molecules of legume seed galactomannan polysaccharides consist of a backbone composed of $\beta(1\rightarrow4)$-linked linear chain of mannopyranose, called the mannan backbone, which is solubilized to a certain extent due to a variable degree of substitution of $\alpha(1\rightarrow6)$-linked single galactose grafts on it. Galactomannans from different legume seeds differ in their mannose-to-galactose (M:G) ratio, molecular weight, and the mode in which the galactose grafts are arranged on the mannan backbone. As thought earlier, the placement of galactose grafts on the galactomannan backbone is not in a regularly spaced sequence. These are placed randomly or in blocks[1,3] on the mannan backbone. Differences in these structural features of galactomannans together determine the functional properties and uses of galactomannans from different plant sources.

Glucomannans are a group of polysaccharides that closely resemble galactomannans. Thus, the glucomannan from the konjac tuber (*Amorphopallus* species) is composed of a linear backbone made up of blocks of $\beta(1\rightarrow4)$-linked mannopyranose units, which are interposed with linear $\beta(1\rightarrow4)$-linked blocks of glucopyranose units. Glucomannans are solubilized due to several O-acetyl groups on some of the backbone of hexose monomers, which act as grafts.[4] Additionally, the glucomannans also have some single glucose grafted on the polymer backbone. Besides konjac tuber, glucomannans are also the main polysaccharide component of aloe vera leaf gel, and since these have mannose in much larger amounts than glucose, they are sometimes referred to as acemannan or acetylated mannans.

4.2   TYPES OF INTERACTIONS OF GALACTOMANNANS

The galactomannan interactions discussed herein are classified as

1. Those with polymeric molecules of its own kind, or other polysaccharides present in a solution, along with it.
2. Superentanglement of galactomannans.
3. High surface activity of fenugreek gum.
4. Interactions with polymeric solids (cellulose and gangue minerals) with hydrophilic surfaces. These interactions find industrial applications in the paper industry and in mineral processing.
5. Regulation of carbohydrate and lipid absorption and metabolism in man and animals.
6. Interactions with low molecular weight substances, where the interactions can also be stereoselective.
7. Moisturizing and self-gelling behavior of aloe vera glucomannan, which has earned it a place in cosmetics and health care products.

Some of these interactions are more of an academic interest, whereas some others have industrial applications. These interactions arise due to the configurational difference between glucose (all *trans*-hydroxyl pairs) polymers on one hand and galactomannan (*cis*-hydroxyl pair in their component sugars) polymers on the other hand, and these are elaborated in the following sections.

4.3   CONFIGURATIONAL DIFFERENCES BETWEEN GLUCOSE POLYMERS AND GALACTOMANNAN POLYMERS

In Figure 4.1 the configurational placement of various hydroxyl groups in the eight possible configurational isomers of D-hexopyranose is shown. Of these, only three—glucose,

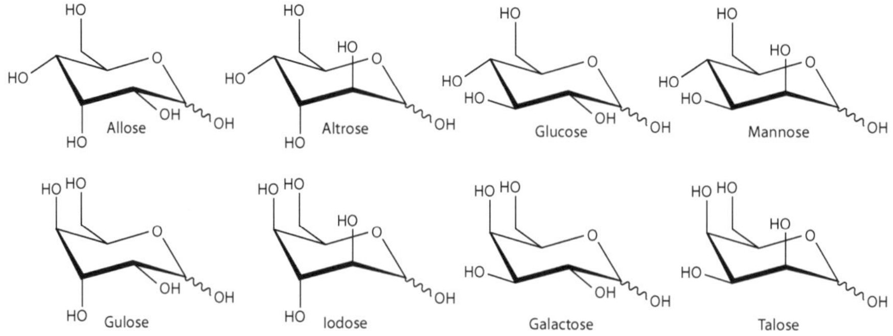

FIGURE 4.1   Configurational placement of hydroxyl groups in the eight isomeric D-hexopyranoses. C-2, 3 OH in mannose and C-3, 4 OH in galactose have *cis*-configuration.

mannose, and galactose—are synthesized in nature and these form the components of several oligosaccharides, polysaccharides, and biomolecules. The remaining three isomers, having more of axially placed hydroxyls, are not preferred during natural biosynthesis because of their higher energy and more of axial interactions.

Glucose is the most common and abundant sugar produced in the plant kingdom. It has the lowest ground-level energy among all the hexoses, due to all its equatorially placed HO groups (no axial repulsion) in its pyranose structure. In plants, glucose and mannose are interconvertible by epimerization, that is, by the change in configuration at C-2. This appears to be taking place in nature when a stronger hydrogen bonding and more moisture-holding capacity is desired in biosynthesized polysaccharides, which are derived from mannose.

In mannose, C-2a, C-3e hydroxyls have a *cis*-configuration, whereas glucose has *trans*-configuration for C-2e, C-3e hydroxyl groups. Though there is very little difference between steric proximity of these hydroxyl pairs of mannose and glucose, considerable differences are realized in some of the reactions and interactions of their β(1→4)-linked polymers (i.e., the glucans (cellulose and its derivatives) on one hand and the galactomannans on the other hand).

There is another group of polysaccharides called glucomannanns. For glucomannanns, having linear, β(1→4)-linked mannan backbone, which is interposed by β(1→4)-linked blocks of glucose monomers, the polysaccharides are solubilized due to the presence of several O-acetyl grafts. Glucomannans have many similarities to galactomannans.

It has been presumed that the differences between cellulose and its soluble derivatives—glucans and galactomannans—arise due to the configurational differences of the component hexose units of these polysaccharides. The presence of a *cis*-pair of hydroxyl groups in C-2, C-3 positions in mannopyranosyl and C-3, C-4 positions in galactopyranosyl units of galactomannans[5] makes a major difference between the functional behavior of glucans and mannans. To a large extent, interactions and the differences between β(1→4)-linked galactomannans and β(1→4)-linked glucans can be explained on the basis of the presence or absence of a *cis*-pair of hydroxyl groups in the component sugars in their molecules.

4.4 HYDROGEN BONDING INTERACTIONS IN CARBOHYDRATES

Extensive literature has been published on the hydrogen bond geometries of carbohydrates.[6] In the molecular species, a hydrogen bond exerts cohesive, electrostatic force over a pretty large distance (2–4 Å). Such electrostatic force gradually diminishes with O···HO distance, which in turn is determined by O···O separation. In case of β(1→4)-linked galactomannans, intramolecular hydrogen bonding is possible between C2-OH, C3-OH groups, having *cis*-conformation. There is a slight difference between C2, C3 *cis* and *trans*, O···O separations. These are ~2.80 Å for *cis* and ~2.86 Å for *trans*, that is for mannose and glucose, respectively. The dihedral angle (HO-C···C-OH), at 2, 3 positions in glucose and mannose are, respectively, ~60° for the *trans*-pair and slightly less (~56°) for the *cis*-pair. This would give rise to a stronger intramolecular hydrogen bond between 2C-OH, 3C-OH of mannose than that in glucose.

In an x-ray crystallographic structure, representing a fragment of cellulose molecule, intrachain hydrogen bonding between C3'-OH and an adjacent pyranoside ring 5-O is generally assumed. When the C3'-equatorial hydroxyls of glucose units in cellulose are involved in such intrachain hydrogen bonding between two adjacent glucose, it results in producing a ribbonlike conformation of linear cellulose molecule in which the pyranosyl sugar rings lie in the plane of a ribbon. Intrachain hydrogen bonding of these ribbon-shaped, linear molecules produces crystallinity in cellulose. Intramolecular hydrogen bonding between *trans*-C2-OH, C3-OH in a glucose monomer of cellulose is less likely because it shall create a situation similar to the formation of a fused five-member ring, which may not be favored due to an increase in axial interaction in the pyranose structure.

By[7] an analogy to cellulose, Wielinga[7] has proposed a similar, three-dimensional hydrogen bonded model for a segment of guar galactomannan, where a C3'-OH to 5-O hydrogen bond is presumed. This is less likely, looking to the observed, higher rotational freedom[1] between two adjacent mannose units in a galactomannan around the pivotal glycoside bond, which preclude or at least reduces the possibility of such bonding. In contrast to a ribbonlike structure of cellulose, $\beta(1\rightarrow4)$-linked mannan chain has been better described as a garland of leaves structure with more rotational freedom at the pivotal point. This explains why the tendency for cellulose to form crystalline regions is far less shared by the mannans. Due to its higher crystallinity, native cellulose only swells in alkali solution, whereas ivory mannan dissolves in it.

In interchain hydrogen bonding to other polysaccharides (e.g., agar, xanthan, and carrageenan) or low molecular weight substances, the C2-OH, C3-OH pair of the ungrafted segment of mannan backbone in a galactomannan can act both as an acceptor or a donor to form a three-center hydrogen bond. "Flip-flop" disorder of a three-center hydrogen bond makes it more durable on a time scale.[8]

We shall now discuss the interactions of galactomannans with other polysaccharides.

4.5 SYNERGIC INCREASE IN VISCOSITY AND GELLING OF GALACTOMANNANS WITH OTHER POLYSACCHARIDES[1]

Certain polysaccharides (e.g., starch, agar, xanthan, and carrageenan) when mixed with guar galactomannan (M:G = ~2:1), results in a synergic increase in viscosity. Formation of a binary gel with improved gel strength takes place upon mixing LBG (M:G = 4:1) with these polysaccharides. In case of fenugreek seed endosperm galactomannan (M:G = 1:1), having the mannan backbone nearly completely grafted by galactose, no synergic increase in viscosity takes place.[9] There is also no synergic change in viscosity of fenugreek polysaccharides on mixing with CMC.

Difference in interchain interaction of these three galactomannans (fenugreek, guar, and LBG), and their interaction with other linear polysaccharides has been attributed to the differences in the size of continuously galactose-grafted mannan backbone (also referred to as the hairy regions) and ungrafted (nonhairy or smooth regions) blocks of the mannan backbone. There exist large (~25) blocks of hairy regions, and even larger blocks (>25) of nonhairy regions in the molecules of certain galactomannans (e.g., LBG). These, hairy and nonhairy regions arise due to continuous placement of galactose grafts on mannan backbone in some regions leaving even larger unsubstituted regions of the backbone in other regions. Thus in LBG, blocks of ~25 continuously galactose grafted mannose units can be present at one site, while leaving portions of even larger ungrafted blocks of the backbone. These nonhairy regions can have strong interchain hydrogen bonding interactions, over an extended length of the backbone with molecules of other linear polymers.

In these interactions, vicinal *cis*-hydroxyl pairs have an important role to play because they can form two- or three-center hydrogen bonds.[6] When hydrogen bonding takes place over an extended portion of a chain, it produces "chain-to-chain zipping" interactions. Multiple hydrogen bonds thus form, have larger energy, and are more durable over an extended period of time, whereas normal hydrogen bonding between small molecules in a solution has much lower energy and has an average bond life of the order of 10^{-12} seconds only. It has been suggested that when there is three-center hydrogen bonding and because of the dynamic nature of the bond, even when one of the bonds snaps off, the participating atoms still remain bonded.

Wielinga and Schumacher[10] have shown a synergic boost in the viscosity of deploymerized and extremely low viscosity (Brookfield viscometer, 10–15 mPas for 1% solution) guar gum on blending with 25% (w/w) Carbopol-820 (trade name of a polyacrylic acid) to several thousand mPas. According to these researchers, the increase in viscosity might be due to "the polymeric acid joining into the matrix of gum thickener."

4.6   SUPERENTANGLEMENT OF GALACTOMANNAN

In a dilute solution, galactomannan molecules exist as mobile, random-coil molecules, yet the rheological properties of galactomannans differ from other random-coil polysaccharides.[11] When a straight line plot of double logarithmic zero-shear specific viscosity (η_{sp}) of galactomannans (guar gum and LGB) against their degree of space occupancy (represented by c[η]) was made, it showed an abrupt change of slope at c[η] ~2.5. Most other random-coil polysaccharides (CMC, alginates, carrageenan, etc.) showed this change only at c[η] ~4.0. In the case of galactomannans, subsequent slope of this plot was also higher being c[η] ~4.5 compared to other random coil polymers having c[η] ~3.3 only. This difference has been attributed to the so-called hyperentanglement, which is a specific characteristic of galactomannan polymer chains.

Let us now consider why random-coil molecules of galactomannans entangle more than other random coil polymers. The change in slope of the plots referred to earlier is supposed to be due to a transition from a dilute or the ideal solution behavior of any random coil polymer sol, that is, when there is no coil–coil interaction, except a normal topological entanglement or coil–coil interaction. In contrast to this, hyperentanglement involves a stronger association of polymer coils, which must be attributed to interaction between nonhairy regions of mannan backbones on different random coil molecules. This is also accompanied by a large increase in solution viscosity.

A few questions now arise: What exactly is meant by strong interaction, and what structural features of galactomannans cause hyperentanglement? Is this due to the presence of a *cis*-pair of hydroxyls on the mannan backbone? The literature currently available is silent on these questions.

First let us consider the insoluble nature and crystalinity of natural cellulose. All the hydroxyls of the glucose units of a cellulose molecule, being equatorially placed, are in the same plane as the pyranose rings. These hydroxyl groups can have good interchain hydrogen bonding, that is, between 3′-OH and 5-O of the adjacent glucose in a molecule, producing a ribbonlike conformation. This is less likely with mannose backbone, as discussed earlier. Interchain hydrogen bonding between parallel and adjacent molecular ribbons of cellulose produce crystalline regions, resulting in a lack of solubility. This is largely prevented in case of galactomannans by frequent galactose grafts on the backbone. Additionally, a "garland of leaves" like polymer chain conformation is likely to have much weaker interchain hydrogen bonding, which shall break on heating and cause solubility.

For cellulose derivatives, the chemically introduced grafts (carboxymethyl, alkyl, and hydroxyalkyl) reduce chain–chain association and induce solubility, but still no hyperentanglement occurs. This is due to the difference in the hydrogen bonding ability of *cis*- and *trans*-pairs of hydroxyls.

We conclude that the association of smooth portions of the galactomannan backbone in random-coil molecules involves a strong, zipping type, three-center hydrogen bonding of the *cis*-hydroxyl pairs, which cause hyperentanglement. For cellulose, long and ribbon-shaped (not random coil) molecules can orient parallel to cellulose fibers and produce crystalline regions. This is not possible in case of soluble cellulose derivatives.

4.7 HIGH SURFACTANT ACTIVITY OF FENUGREEK GALACTOMANNAN[12,13]

Soluble polysaccharides owe their emulsion stabilizing ability to the high viscosity that they impart to a sol rather than due to any significant reduction in interfacial and surface tension. In contrast to this, fenugreek galactomannan shows a unique property of reducing interfacial tension and surface tension in a sol, which is comparable to that of gum acacia. Gum acacia owes its high surface activity to its being a composite of a hydrophobic protein and a hydrophilic polysaccharide fraction, which is a condition necessary for surfactant activity of any substance. Purified fenugreek polysaccharide, which is completely protein free, still reduces the surface tension at a water–air interface by as much as >30 mN/m, which is much larger compared to any other galactomannans. Why is this so?

Fenugreek galactomannan has a M:G ~1:1, which means that its mannan backbone has become less hydrophilic having been completely shielded by galactose grafts. Absence of nonhairy blocks on fenugreek mannan backbone does not allow it to have any zipping-type hydrogen bonding interaction with molecules of its own kind. Rigid rodlike molecules of fenugreek galactomannan has a tendency to migrate to a water–air interface to reduce the surface tension.

When a mineral or a vegetable oil is dispersed in an aqueous fenugreek galactomannan sol, the gum gets adsorbed on the surface of the tiny oil droplets. Upon stirring, oil gets emulsified due to the polysaccharide adsorbed on their surface, and these emulsified oil droplets are prevented from coalescence and flocculation.

Adsorption of gum on the surface of emulsified oil droplets can be attributed to the *cis*-OH in galactose grafts, which being less hindered and being removed from the manna backbone, are still able interact via hydrogen bonding to the oxygen functions in the oil to form a surface film and prevent coalescence.

4.8   ADSORPTION OF GALACTOMANNANS ON HYDROPHILIC SOLIDS: APPLICATIONS IN THE PAPER INDUSTRY[14]

Because of the numerous hydroxyl groups present in their molecules, galactomannans exist in extensively intra- and intermolecularly hydrogen-bonded form, and they strongly interact with hydrogen bonding solvents, such as water and even hydrophilic solids (e.g., cellulose fibers).

In their application as a binder for cellulose fibers in the paper manufacturing, galactomannan polysaccharides act as a better binding material than starch, which is a glucose polymer. This difference is best explained by taking into account the difference in the configuration of C2, C3-OH groups in the repeat units of starch (glucans, *trans*) and a galactomannan, which has a *cis*-configuration.

Galactomannans have the amazing power to adhere to the surface of cellulose fibers in wood pulp by hydrogen bonding. Galactomannans thus adsorbed on cellulose fibers help in fiber–fiber bonding during paper manufacturing. Linear galactomannan molecules, by bridging and cementing cellulose fiber together into a paper sheet, results in an increased strength, while increased recovery of pulp takes place due to aggregation of cellulose fines, which are otherwise lost in the filtrate of pulp liquor.

4.9   GALACTOMANNANS AS A DEPRESSANT/FLOCCULENT IN NONFERROUS MINERAL BENEFICIATION

In mineral beneficiation of nonferrous metallic ores (Cu, Ni, Co, etc.), the hydrophilic surface of gangue minerals (impurities of silica, silicates, and clay minerals) adsorb galactomannan on their surface, thereby making these particles even more hydrophilic.[14,15] These are then aggregated and get depressed, while the sulfide mineral particles, having their surface rendered hydrophobic by a floatation agent (alkyl xanthate), float along with the froth. Galactomannans have proved to be more effective depressants than starch dextrins and CMC. This has been attributed to stronger hydrogen bonds formed by the *cis*-hydroxyl pair of long and linear galactomannan molecules over a large surface area of gangue particles, and due to bridging between several gangue particles causing their agglomeration into still larger particles. Thus, there is a more effective separation of sulfide mineral ore from gangue minerals in a froth flotation process using galactomannan as depressants.

4.10   MODULATION OF CARBOHYDRATE AND LIPID METABOLISM[16]

When taken orally in food, the galactomannan-based soluble dietary fiber slows the absorption of sugars and lipids from the intestinal tract into the body fluids. This helps in controlling blood glucose levels in healthy and diabetic persons. The slowing of fat absorption controls blood lipid level (low-density lipoprotein [LDL], cholesterol, and triglycerides), which helps to prevent cardiovascular diseases. Synergic enhancement of the viscosity of food starches and proteins in the digestive system by hydrogen bonding of these food constituents to galactomannan and entrapment of digestive enzymes into viscous gum appears to play a role in these beneficial actions of galactomannan as a food fiber. Nongalactomannan food fibers do exert a similar effect, but weight for weight the effect of a galactomannan is more than any other dietary fiber. Even depolymerized and low molecular weight galactomannans are effective as dietary fiber,[7] which indicates that besides the viscosity of a galactomannan polymer, its tendency to interact with other polymers has an important role to play.

4.11 FORMATION OF FUSED, FIVE-MEMBER RING BY *CIS*-PAIR OF HYDROXYL GROUPS IN GALACTOMANNANS

Following are some other examples of reactions and interactions of galactomannans and smaller molecular species.

It is a well-established fact that cyclic-acetonides, cyclic-carbonate esters, and cyclic-borate complexes at C-2, C-3 positions in mannopyranose are formed, because the hydroxyl groups at these sites are in a *cis*-configuration (Figure 4.2). The formation of these fused five-member rings is dependent upon the overlap of the van der Waal's radii of rather small atoms (C, O, and B) of the second group of periodic table. Glucopyranose in cellulose does not form fused, five-membered ring the at the *trans*-, C-2, C-3 positions.

In spite of little difference in the steric placement of hydroxyl groups, the ring formation ability of *cis*- and *trans*-hydroxyl groups are quite different.[17] This is explained on the basis of deformation of the sugar pyranose ring due to the formation of a

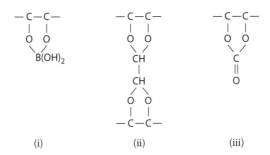

FIGURE 4.2 Formation of five-membered cyclic ring by the 2,3 *cis*-hydroxyl pair of mannose backbone in galactomannans. (a) Borate ester. Two free hydroxyl groups of the ester can hydrogen bond to another mannan chain at pH >7 to cross-link, resulting in gelling or insolubilization. (b) Cross-linking, resulting in insolubilizing due to di-acetal formation with glyoxal at pH <4. The reaction is reversed at pH >7. (c) Cyclic carbonate ester formed by the reaction of phosgene.

slightly flattened and fused five-member ring. Flattened ring formation at the *cis*-2a, 3e hydroxyl pair is accompanied by a reduced interaction between axial hydrogen atoms in the pyranose ring, which results in favoring such fused-ring formation. In contrast, when such an event happening with *trans*-2e, 3e hydroxyl pair shall push the axial groups inward and increases their interaction, so that the ring formation is not favored.

In the oxidation reactions of *vic*-glycol group, involving lead tetraacetate, sodium periodate, potassium permanganate and osmium tetraoxide, five-member cyclic intermediates are presumed to be formed by both *cis*- and *trans*-glycols. However, the rate of oxidation reaction with *trans*-diol is slower. If these oxidants do really form five-member cyclic intermediates with both *cis*- and *trans*-isomers, it is possible that five-membered ring intermediates can be formed, even with a *trans*-isomer, when a large-sized heteroatom (Pb, I, Mn, or Os) is involved. The fused five-membered rings with a large heteroatom are likely to be less strained compared to those having smaller-sized atoms, that is, C, O, and B.

4.12 INTERACTION WITH CROSS-LINKING AGENTS AND ITS APPLICATIONS[18]

Cross-linking with borate and glyoxal, discussed earlier, cause guar gum and other galactomannan gums to form a gel, or at very low concentrations of cross-linking, render them water dispersible and slow hydrating. Such cross-linking involves hydrogen bonding and complexing of the borate ion between two different mannan chains, via pairs of *cis*-OH. Being pH dependent, such cross-linking can be reversed by the change of pH. This has been used for making galactomannan formulations needing delayed viscosity development. Borate cross-linked guar gum formulations have found extensive industrial applications, particularly in oil field drillings. The mechanism of cross-linking of guar by borate ion [$B(OH)_4^{-1}$] involves complexing/hydrogen bonding at alkaline pH (>7). In contrast, glyoxal cross-linking takes place at acidic pH (~3–4), and involves bis-acetal (five-member rings) formation with two *cis*-hydroxyl pairs on different chains. The pH requirements, in these two cross-linking reactions are complementary. These reactions are characteristic of polysaccharides containing mannose backbone, while nonmannose polysaccharides either do not show these interactions, or these are very weak. Controlled cross-linking with borate and glyoxal has been used in production of guar gum products having pH-dependent and delayed viscosity development formulation. Borate cross-linked guar gum disperses at pH~9 and develops viscosity only when pH is lowered to <7. In contrast, glyoxal cross-linked dispersible products, which are formulated at pH~3–4, disperse only when the pH is raised to ~7–8 by addition of alkali.

4.13 CROSS-LINKING WITH TRANSITION METAL IONS[19]

Complexing of certain metal ions—typically Cu(II), Zr(IV), Sb(III), Ti(IV) and CrO_4^{-2}—by *cis*-hydroxyl pair can result in controlled gelling or viscosity enhancement of guar gum paste. According to Dea,[1] these interactions arise due to hydrogen bonding associations involving the *cis*-pair of hydroxyl groups. Such hydrogen bonding associations are of a higher energy and longer duration compared to normal hydrogen bonds. Such interactions have been extensively applied in oil-well drilling in the petroleum industry, and these interactions have also been used for enhanced oil recovery by fracturing of an oil well. Complexing involving *cis*-OH is reduced by derivatizing guar gum to its hydroxypropyl ether (HP-guar gum) due to the blocking of some of the *vic*-hydroxyl groups. In case of a double derivative, CM, HP-guar gum, carboxymethyl group is also involved in complexing reaction with transition metal ions. These derivatives are thermally more stable than natural guar gum.

4.14   INTERACTION WITH RACEMIC MIXTURES INVOLVING CHIRAL-SELECTIVE INTERACTIONS OF GALACTOMANNANS

Cellulose and starch have frequently been used as column material in chromatographic separation of organic compounds. These polysaccharides are composed of optically active sugar (glucose) monomers, but they do not exhibit high chiral selectivity for racemates.[20] In contrast to this we have observed good chiral selectivity during chromatography on galactomannan polysaccharide (guar gum) as an adsorbent toward certain groups of racemates, particularly the amino acids. Enantiomers in certain racemic mixtures show chiral-selective interaction in hydrogen bonding to the *cis*-hydroxyl group pair on the mannopyranose ring. This principle has been used by this author to develop *chiral chromatography* and enantiomer separation, using galactomannan as an adsorbent.

According to observations by myself and colleagues, 2a, 3e *cis*-hydroxyl groups of mannan backbone in guar galactomannan during hydrogen bonding interaction with an external donor exhibit enantioselectivity. This has been observed in the resolution of racemic amino acid mixture. α-Amino acids have an chiral carbon atom linked directly to the amino and carboxyl groups, and they show good chiral selectivity in forming simultaneous hydrogen bonds via –NH$_2$ and –COOH groups with vicinal hydroxyls in a *cis*-configuration. This can be demonstrated from scale models of such molecules.[21,22]

In chromatography using a nonchiral media (silica gel), optically active 1, 2-diphenylethane-1, 2-diol, has been used as a *chiral mobile phase additive* for the separation of certain racemates. Because of the bulkiness of the phenyl groups, *syn*-conformation of the hydroxyl groups in this molecule is stabilized by restricted rotation at the Φ-C-C-Φ bond. Preferred *syn*-conformation in such a diol is known to interact selectively in hydrogen bonding to chiral isomers of a racemic mixture resulting in their separation.[23]

A similar condition is created in the steric placement of the *cis*-pair of hydroxyl groups on a mannopyranose ring system, and these are similar to the *syn*-configuration of optically active 1, 2-diphenylethane-1, 2-diol. This explains the chiral selectivity of guar galactomannan polysaccharide toward certain sets of racemates, which can form two simultaneous hydrogen bonds with molecules, having α-amino acid type placement of an atom, near the chiral carbon atom. More work on these lines needs to be done.

4.15 MOISTURIZING ACTION AND GELLING OF GLUCOMANNAN POLYSACCHARIDE IN ALOE VERA PLANT[24]

Leaf extract containing glucomannan polysaccharide in the mucilaginous gel of the thick epidermis of the aloe vera plant (*Aloe barbadenis Miller*) got a renewed interest from the cosmetic and pharmaceutical industries during the past two to three decades. This has been attributed to the soothing action of aloe extract on skin and other therapeutic values of glucomannan in the extract. Glucomannan polysaccharides form the major solid constituents, which is >0.6% of a total of ~1% solids in the extract of this gel, the rest (~99%) being water. These polysaccharides are considered innocuous for topical applications and oral use.

Structural study[25] of the major aloe vera gel polysaccharide has revealed it to be a water-soluble, linear, $\beta(1\rightarrow4)$-linked acetylated glucomannan of high molecular weight (>2 × 10^5 Dalton). The degree of substitution of acetyl group (DS ~0.6–0.85) in aloe vera glucomannan is rather high, and the acetyl substituents are nonuniformly placed on the mannan backbone. Some hexose monomers in these polysaccharides have been reported to carry two to three acetyl substituents, which can render those monomers less hydrophilic. aloe vera glucomannan owes its water solubility to these acetyl grafts, which reduce chain–chain interaction of the glucomannan backbone. These random coil molecules do not have any tertiary structures; yet in the native form, aloe vera polysaccharides occur as a transparent, gelatin-like gel in the leaves. Of the various isolated and characterized aloe vera polysaccharides, the fraction rich in acetyl content is reported to form a gel similar to that occurring in the freshly cut aloe vera leaf. On deacetylation, gelling property of this fraction is lost.[26]

Native aloe vera gel, from freshly cut epidermis of the leaf, show syneresis and it is thixotropic. On shearing, aloe vera gel, irreversibly changes into a fluid. Commercial aloe vera extracts are sold as fluid. Powder and binary aloe vera gels are produced by adding it to Carbopol (polyacrylic acid polymer), or microbial and seaweed polysaccharides, xanthan and carrageenan, respectively.

Polysaccharides having tertiary structures (e.g., agar and carrageenan) are self-gelling at an overall low solid concentration (1%–2%). Polysaccharide gelling in a binary mixture at a much lower concentration can involve either association, that is, a mixed polymer complex formation or segregation of noncompatible polymers with microphase separation.[27]

What makes the acetyl-rich aloe polysaccharide, having no tertiary structure, self-gelling at such an extremely low (~0.6%) concentration? There are two plausible explanations:

1. A segment (or several segments) of aloe vera polysaccharide molecules, which is heavily substituted by acetyl groups becomes hydrophobic and tends to segregate from the hydrophilic segment of its polymer chain. Thermodynamically, this is possible in a large and linear macromolecule. Incompatibility between hydrophobic and hydrophilic segments of this liner chain polymer results in a microphase separation. Strongly hydrated, hydrophilic microphase is entrapped in an outer hydrophobic phase,

resulting in gelling. Shearing disturbs the outer hydrophobic phase, resulting in syneresis and thixotropic rheology.

2. Besides the soluble glucomannan, there is a report of the presence of an
   insoluble, cellulose fraction in aloe vera polysaccharides, which is similar to the commercial, microcrystalline cellulose (MCC).[26] It is, therefore, possible that native gel in aloe vera leaf is a sort of composite or
   a hydrocolloid-alloy type of gel, similar to those that are now commercially produced from MCC and soluble gums (e.g., guar galactomannan
   or CMC).[28]

4.16   CONCLUSION

There is a general agreement that hydrogen bonding interactions of galactomannan
and glucomannan polymers are much stronger compared to most of the other die-
quatorially, (β-1, 4)-linked linear polysaccharides (e.g., cellulose). Yet, there appears
to be some hesitation in proposing a stronger and time durable, three-center hydrogen bonding, arising in case of these mannan polysaccharides, due to the presence
of C-2, 3-hydroxyls, with a *cis*-configuration. Our own suggestions are not based on
much new experimental data, yet the chiral selectivity of 2a, 3e hydroxyl groups in
mannose, first observed by myself and colleagues, certainly points to the fact, that
these hydroxyl pairs do create a major difference in hydrogen bonding reactions
and interactions of mannans when compared to other (β-1, 4)-linked glycans (e.g.,
cellulose).

REFERENCES

1. Dea, I. C. M., Structure/function relationships of galactomannans and food grade cellulosics, In Gums and Stabilizers for the Food Industry 5, G. O. Phillips, P. A. Williams,
   and D. J. Wedlock, eds., IRL Press at Oxford University Press, Oxford, 1990, 373–382.
2. Dea, I. C. M. and Morrison, A., Chemistry and interactions of seed galactomannans,
   Adv. Cabohydr. Chem. Biochem., 31 (1975): 241–312.
3. Mathur, N. K. and Mathur, V., Fenugreek and other lesser-known legume galactomannan-
   polysaccharides: Scope for developments, J. Sci. Industr. Res., 64 (2005): 475–481.
4. Kato, K. & Matsuda, K., Chemical structure of konjac glucomannan, Agri. Biol. Chem.,
   33 (1969): 1446–1453.
5. Mathur, N. K. Mathur, V., and Nagori, B. P., Fine structure of galactomannans
   and glucomannans and correlation to their interactions, In Trends in Carbohydrate
   Chemistry, Vol. 3, P. L. Soni, ed., Surya International Publications, Dehradun, India,
   1997, 119–123.
6. Jeffrey, G. A. and Saenger, W., Hydrogen Bonding in Biological Structures, Springer-
   Verlag, Berlin, 1991.
7. Wielinga, W. C., Production and applications of seed gums, In Gums and Stabilizers for
   the Food Industry 5, G. O. Phillips, P. A. Williams, and D. J. Wedlock, eds., IRL Press
   at Oxford University Press, Oxford, 1990, 383–403.
8. Steiner, T. and Saenger, W., Reliability of assigning O---H···O hydrogen bonds to short
   intermolecular O···O separations in cyclodextrin and oligosaccharide crystal structures,
   Carbohydr. Res., 259 (1994): 1–12.
9. Garti, N. and Pinthus, E. J., Fenugreek gum: The magic fiber for an improved glucose
   response and cholesterol reduction, Nutracos, May/June (2002): 5–10.

10. Wielinga, W. C. and Schumacher, G., European Patent Application No. 81304321.3, (2001) Publication number 0 048 612/ A2, European Patent Office.

11. Morris, E. R. Cutler, A. N. Ross-Murphy, S. B., and Rees, D. A., Concentration and shear rate dependence of viscosity in random coil polysaccharide solutions, Carbohydr. Polym., 1 (1981): 5–21.

12. Garti, N., Surface activity of fenugreek seed gum, J. Dispers. Sci. Tech., 20 (1-2) (1990): 327–350.

13. Garti, N. and Reichman, D., Fenugreek seed endosperm galactomannan, Food Hydrocolloids 8(2) (1994): 155–173.

14. Whistler, R. L. and Hymowitz, T., Guar: Agronomy, Production, Industrial Use and Nutrition, Purdue University Press, West Lafayette, IN, 1979, 101–113.

15. Somasundram, P. and Mudgal, B. M., Reagents in Mineral Technology, Marcel Dekker, New York, 1967.

16. Morris, E. R., Physicochemical properties of food polysaccharides. In Dietary Fibers: A Component of Food, T. F. Schweizer and C. A. Edward, eds., ILSI Human Nutrition Reviews, Springer-Verlag, London, 1992, 41–56.

17. Nasipuri, D., Stereochemistry of Organic Compounds, 2nd ed., Wiley Eastern Ltd., New Delhi, 1993, 244.

18. Maier, H., Anderson, M., Karl, C., Magnuson, K., and Whistler, R. L. Guar, locust bean, tara and fenugreek gums. In Industrial Gums, 3rd ed., R. L. Whistler and J. N. BeMiller, eds., Academic Press, New York, 1993, 181–226.

19. Henkel Corporation, Guar and Derivatives: Oilfield Applications. Technical Bulletin, Houston, Texas, 1986.

20. Mathur, R., Bohra, S., Mathur, V., Narang, C. K., and Mathur, N. K., Chiral ligand-exchange chromatography on polygalactomannan (guaran), Chromatographia, 33 (1992): 336–338.

21. Mathur, R., Bohra, S., Mathur, V., Narang, C. K., and Mathur, N. K., Guaran: A novel polysaccharide for racemate resolution, J. Liq. Chromatogr., 15 (1992): 573–584.

22. Mathur, V., Kanoongo, N., Mathur, R., Narang, C. K., and Mathur, N. K., Resolution of amino acid racemates on borate-gelled guaran-impregnated silica gel thin layer chromatographic plates, J. Chromatogr. 685 (1994): 360–364.

23. Allenmark, S. G., Chromatographic Enantioseparations: Methods and Applications. Ellis Harwood Ltd., Chichester, 1988.

24. Whistler, R. L., and BeMiller J. N., eds., Industrial Gums, 3rd ed., Academic Press, New York, 1993, 228–230.

25. Gowda, D. C., Neelisiddaiah, B., and Anjaneylu, Y. V., Structural studies of polysaccharides from aloe vera, Carbohydr. Res., 72 (1979): 201–205.

26. Gowda, D. C., Structural studies of polysaccharides from Aloe saponaria and Aloe vanbalenii, Carbohydr. Res., 83 (1980): 402–405.

27. Piculell, L., Iliopoulos, I., Linse, P., Nilsson, S., Turquois, T., Viebke, C., and Zhang, W., Association and segregation in ternary polymer solutions and gels. In Gums and Stabilizers for Food Industry 7, G. O. Phillips, P. A. Williams, and D. J. Wedlock, eds., Oxford University Press, Oxford, 1993, pp. 309–322.

28. McKinley, E. J. and Tauson, D. C., Applications of microcrystalline cellulose. In Gums and Stabilizers for Food Industry 7, G. O. Phillips, P. A. Williams, and D. J. Wedlock, eds., Oxford University Press, Oxford, 1993, 309–322.

5 Rheology of Hydrocolloids

5.1 INTRODUCTION

Applications of plant gums in food as well as in nonfood industries are always made in aqueous media. On putting in water, the molecules of hydrocolloids bind water to varying extant, and dramatic changes take place in their size, shape, and molecular conformation. Hydrocolloids in an aqueous media act as modifiers of the functional behavior of liquid water. Changes in the functional behavior of water can be observed as thickening and gelling, or a behavior intermediate between these two. Though these changes can be observed manually and visually, there is a need for a quantitative evaluation of such changes. One important property of fluids is the flow behavior, which changes due to thickening. Thickening of a sol is a measurable property in terms of the viscosity of a fluid.

Gels do not flow, but they are elastic and are deformed when stress is applied to them. The deformation of gels can also be quantitatively evaluated. Quantitative study of the flow behavior of fluids and deformation of soft gels is called *rheological studies* and this has now become a well-developed field of applied sciences. For simplicity, the subject of rheology has been dealt here only in a nonmathematical and qualitative way.

5.2 RHEOLOGY OF HYDROCOLLOID SOLS

Aqueous polysaccharide systems are of interest to a wide range of scientists, technologists, traders, and consumers because of their very special rheological properties. Since the present book is meant for scientific and technical personnel concerned with the study, manufacture, development, and applications of galactomannan-based hydrocolloids or gums, it is necessary for them to have an elementary knowledge of the science of rheology. This chapter is meant to impart a preliminary knowledge of rheology to the readers. Rheology is a well-developed branch of science, with strong mathematical footings, but for practical purposes, it is only necessary to know the qualitative aspects of the subject, and this shall be dealt with here.

The term *rheology* can be defined as "the study of deformation and flow of any material, and it embraces such properties of a material as elasticity, plasticity and viscosity".[1,2]

Polysaccharides being hydrocolloids or gums have the following two most important behavioral properties when they are present in an aqueous system:[3]

1. Polysaccharides thicken aqueous systems.
2. Some polysaccharides, under certain specific conditions, can also form gels in an aqueous system.

There are also some polysaccharide-based hydrocolloids that exhibit a mixed thickening and gelling behavior. Such hydrocolloids are referred to as having a mixed sol-gel rheology. Overall, the thickening and gelling of an aqueous system are the two rheological properties of particular interest to polysaccharide users or more generally any hydrocolloid user. In scientific terms thickening is referred to as an increase in the viscosity of a fluid system.

5.3   VISCOSITY DEFINED[4]

Viscosity can be visualized as the tendency of a fluid to resist flow. Every fluid, whether it is a pure liquid, a solution, a colloidal sol or a gas, has a finite viscosity. In the case of a solution, on increasing the concentration of a dissolved substance or increasing a dispersed phase in a colloidal sol, generally there is an increase in viscosity. This is observed as a thickening of the fluid system. The viscosity enhancement is even more with high molecular weight solutes, particularly the linear polymers, which have high volume occupancy.

Viscosity in any fluid arises due to the internal friction between the layers of a fluid, gliding past each other and the friction arises due to movement of one layer of fluid relative to the other. The force required to move a fluid layer is greater when the friction is more, and this movement is called shearing, and the force is called *shearing force*. More viscous fluids need higher shearing force, compared to less viscous fluids.

The term *fluidity* is the reciprocal of the viscosity, which is denoted by η. Hence, fluidity equals $1/\eta$.

When a force is applied to a volume of elastic material, then a deformation occurs. We imagine two planes, each of area (A), which are separated by a small distance (dx). Now one plane is moved relative to the other, at a velocity (dv) by applying a force (F). According to Newton's law, the *shear stress*, that is, the force divided by area, parallel to the force shall be, F/A, and this is proportional to the *shear strain rate* or the velocity gradient (dv/dx), that is, the rate of change of velocity and the distance between the sliding plates of liquid. The proportionality constant (η) is known as the *dynamic viscosity*. Thus, $F/A = \eta\, dv/dx$. Hence, $\eta = (F/A)/(dv/dx)$, or Shear stress/Shear strain.

Shear strain is the rate of flow of a fluid or the rate of its stirring. It is quantified by the displacement (D) per unit height (H), that is, D/H, and the rate of this effect, shear strain rate is the velocity (V) per unit height (V/H), where the height is the distance from a relatively unaffected (stationary) position of fluid layer.

The SI unit for viscosity is pascal (Pa). When a shear stress of one dyne/cm^2, produces a shear rate of one reciprocal second, the viscosity shall be one pascal. For practical purpose, viscosity is also expressed in centipoises per second (cps). One cps = one millipascal, or one pascal = 10 cps.

The shear strain is quantified by the displacement per unit height and the rate of this effect is the velocity gradient per unit height (dv/dx), where the height is the distance to a relatively unaffected position.

5.4 NEWTONIAN AND NON-NEWTONIAN FLUIDS: PSEUDOPLASTICITY[1,2]

For pure water and aqueous solutions containing dissolved, low molecular weight material, the viscosity is independent of shear strain rate. In such cases the plot of shear strain rate (i.e., the rate of stirring) plotted against shear stress is linear and passes through the origin, as shown in Figure 5.1. Such systems are referred to as *Newtonian fluids*. Increasing concentration of a dissolved or dispersed substance generally causes thickening of a solution, resulting in an increase in the viscosity. Increase in viscosity is higher for high molecular, linear polymers.

Those fluids for which the viscosity changes with shear strain are termed *non-Newtonian fluids*. Typical examples of non-Newtonian fluids are the solutions of polymeric hydrocolloids. Non-Newtonian fluids are termed *pseudoplastic* or *shear thinning* materials, for which there is an instantaneous decrease in viscosity with increase in shear strain. At concentrations higher than a *critical value*, most of the hydrocolloid solutions exhibit non-Newtonian behavior where their viscosity depends on the shear strain rate. Most of the polysaccharide solutions behave as non-Newtonian fluids, and being shear thinning, these fluids are easier to pump and mix because they become thinner on shearing.

Pseudoplasticity of polysaccharide sols arises due to the linear and high molecular weight molecules getting untangled from each other and getting oriented in the direction of flow of the sol. Generally this behavior increases with the concentration of a polymer in a solution.

Low viscosity fluids generally exhibit flow characteristics that are Newtonian, and in such cases there is a linear relationship between the applied shear stress and the shear rate. When a plot of shear stress *versus* shear rate is made, it comes to be a straight line, the slope of which is equal to the viscosity of the fluid.

Most of the polysaccharides gums exhibit non-Newtonian flow behavior when present as viscous sol. Sol systems that do not have yield value and do not show a linear relationship between shear stress and shear rate are termed pseudoplastic

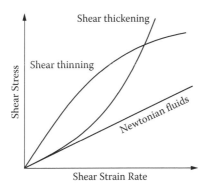

FIGURE 5.1 Classification of fluids, according to viscosity behavior.

materials. These exhibit shear thinning, meaning that as the shear increases, the fluid becomes thinner and the flow rate increases.

Some polysaccharide sols do not flow immediately upon applying a shearing force, but they begin to flow after a limiting initial force (which is called the *yield value*) is applied. Such fluids are referred to as *thixotropic sols*. In the case of a thixotropic polysaccharide sol system, the *yield point* is defined as the minimum shear force required to cause the transition of gum dispersion, appearing as a highly elastic gel into a *flowable sol*. When the shear rate is low, such a sol looks like a viscous semisolid. This semisolid thins out at higher shear rates and becomes flowable. A gum sol in such a case is presumed to be a very soft gel with a well-defined polymer network structure when no shear is applied. When an increasing shearing force is applied to this semigel, it flexes until the polymer network is disrupted and true flow begins. On removing the shear, the system starts rebuilding of the gel network again. This can happen either immediately or over some time. This network allows a relatively thick cluster of water to remain on the surface of the hydrocolloid particle to which it clings tenaciously.

5.5 THIXOTROPIC FLUIDS[2]

Thixotropic fluids exhibit a time-dependent response to shear strain rate over a longer period than that associated with changes in the shear strain rate. They may liquefy when being stirred or shaken and then again solidify, or may not solidify, when shearing is stopped.

The term *dilatancy* or *shear thickening* is used for fluids that show an increase in viscosity with shear stress and strain due to structural enhancement. Though such sols are uncommon, a typical example is that of uncooked cornstarch paste, where shear stress squeezes out the water from in-between starch granules, allowing them to grind against each other. This property is often utilized in sauces (e.g., tomato ketchup), where its flow from a bottle does not take place at low shear stress. When shaking of the bottle increases shear stress, an increased flow can be observed. Dilatancy results due to an increase in the volume of suspensions of irregular particles in a fluid, which results in the creation of small cavities between the particles as they scrape past one another.

5.6 HYDRATION AND SOLUBILITY OF POLYSACCHARIDES

For polysaccharides we are concerned with the term *rheology* when referring to their aqueous dispersions or the colloidal sols. An extensive change of the conformational structure of polysaccharides, from their normal molecular conformations in the solid state to supramolecular structures, takes place in an aqueous solution and gel system. Because of the hydration of polysaccharide molecules in an aqueous medium, its solution exhibits an increase in viscosity, and gelling can result in some cases.

In an aqueous medium, hydration of a polar molecule or an ion always precedes its dissolution. The term *hydration*, which means "binding to water," is difficult to define and it is not fully understood. Water by itself is a very complex material composed of highly polar molecules. Hydration, which results due to ion–dipole or dipole–dipole interaction, is called hydrogen bonding because one of interacting species during hydration is a dipolar water molecule. The mutual effects of water on polysaccharide and polysaccharide on water are extremely complex. These become even more complex in the presence of other water-soluble materials present in a system, such as sugar and salts, which are common components of most food materials.

In an aqueous system, water strongly competes for hydrogen bonding to those sites, which are already intra- and intermolecularly hydrogen bonded in polysaccharide molecules. The energy of hydrogen bond (1–10 cal) is variable and in most of the cases hydrogen bonding is dynamic in nature. Hydrogen bonding determines the flexibility and the conformation, which a polysaccharide molecule acquires when it is fully hydrated prior to its dissolving.

It was earlier mentioned that it is difficult to explain the term *bound water* in a system, which is sometimes referred to as *nonbulk water* to differentiate it from ordinary water. In contrast to bound water, *unbound water* or *bulk water* is that which can be easily removed from an aqueous system by filtration or centrifugation. Unbound water is even more effectively removed by freeze-drying (i.e., evaporation at extremely low pressure). In the case of carbohydrates, and more particularly for polysaccharides, bound water is further classified as being capable of freezing type or nonfreezing type. According to the freezing behavior, unbound water is further classified in two categories: trapped water and bulk water. Pure water, at the normal atmospheric pressure, freezes at a temperature equal to or below 0°C, which depends on the rate of cooling. Some water may take a much longer time to freeze than others.

Bound water is further classified into *tightly bound water*, which can only be removed by freeze-drying, whereas *loosely bound water* can be removed by filtration or centrifugation. *Bound freezable water* freezes at a temperature that is lower than normal water, that is, it is easily supercooled and it exhibits a reduced enthalpy (latent heat) of fusion or melting. Inability of water to freeze is used to determine water bound to a soluble polysaccharide. In the case of a frozen food (e.g., ice cream) the desirable nonicy texture arises due to bound but nonfreezable water.

5.7 DEPENDENCE OF VISCOSITY OF POLYMERIC MOLECULES ON THEIR SIZE, SHAPE, AND CONFORMATION

For linear polymeric hydrocolloids, the size, shape, and orientation of their molecules with respect to the direction of flow are the major factors that determine the viscosity pattern. In contrast to this, the size and orientation of compact and spheroid molecules (e.g., those of gum acacia) do not have much affect on their flow and hence the viscosity change is far less with shear strain rate.

Linear or substantially linear polymers behave in a qualitatively predictable way with respect to the relationship of their viscosity, and their structure and conformation. In dilute solutions this relationship effectively depends on their *swept out volume* or the *hydrodynamic volume* of the molecules as they tumble and are swept in a solution. At very low concentrations, where there is no effective interaction between different molecules of a polymer, and when they are in their maximum extended conformation, their viscosity is only slightly higher than that of the pure solvent, which is water. This small difference depends upon the total *spherical volume*, which is dependent on the concentration and the radius of gyration of solute molecules, which corresponds to the freely rotating molecules.

The viscosity of a substantially linear polymeric hydrocolloid depends on its cross-sectional area in the direction of flow. At low flow rates, linear molecules with preferred long and thin conformations still have effectively large cross-sections due to tumbling of their molecules in a solution. At a high shear strain rate, such linear molecules tend to get aligned with the direction of flow and have much smaller effective cross-sections. This is because of a lack of tumbling at increased flow rate and hence they have much lower viscosities at a higher shear strain. More compact molecules or spheroids are not so much affected by their orientation, relative to the direction of flow and hence their viscosity changes are minimal with shear strain rate.

The relationship between viscosity and the concentration of a hydrocolloid is generally linear at low viscosity values (~10 millipoise or cps). This means that more extended and linear polymers (e.g., those of galactomannans and soluble cellulose derivatives) increase the viscosity to a greater extent at low concentrations than more compact molecules (e.g., gum acacia) having a similar molecular weight range.

Generally a less flexible linkage between the sequential sugar monomers in a linear polysaccharide chain gives rise to a more extended structure. The linkage spacing, its direction, and charge density are other factors that determine the viscosity of a polysaccharide sol. Polysaccharide molecules, which are most capable of an extended structure, are those having $\beta(1\rightarrow4)$-linked or bis-equatorially linked pyranose sugar monomers. Examples of such polysaccharide molecules are soluble cellulose derivatives, galactomannans, and xylanes. Polysaccharides, which are less capable of extended structure, are those having $\alpha(1\rightarrow3)$ and $\alpha(1\rightarrow4)$ or the bis-axially linked pyranose sugar units (e.g., the amylose and amylopectin fractions of starch and dextran polymers).

At very low concentrations (<0.1%), a linear hydrocolloid, which can exist in a random coil conformation in a solution, is more likely to have Newtonian viscosity. At higher concentrations (>0.5%), such linear polymeric material of a high molecular weight, when present in a sol, starts exhibiting non-Newtonian behavior, and the viscosities become dependent on the shear strain rate. At intermediate concentrations (~0.2%), and above a critical value, most linear hydrocolloid solutions generally exhibit non-Newtonian flow behavior, where their viscosity becomes dependent on the shear strain rate.

5.8 EFFECT OF CHARGE ON THE VISCOSITY OF LINEAR POLYMER

It was been mentioned earlier that uncharged, linear, nonionic polysaccharide molecules in a solution have a tendency to acquire a folded or a random-coil conformation. At a low concentration, these hydrated random coil molecules disperse uniformly throughout the solution. Such a dilute solution obeys the normal laws of dilute solutions, or exhibits the usual colligative properties. There is very little molecular, coil–coil interaction in such cases.

When a shearing force is applied, the random coil molecule can open up to an extent, which depends upon the shear rate. The sol can then exhibit Newtonian or non-Newtonian rheology, depending upon the viscosity that it ultimately develops. For a charged, linear polysaccharide, (e.g., carboxymethylcellulose [CMC], sodium alginate, or carboxymethylated guar gum) charge repulsion taking place on different portions of the polymer chain, to a large extent, prevents a tight, random-coil formation. Hence, the molecular volume occupancy in such cases is much more. This is particularly true of polymers, where the anionic charge is distributed uniformly along the total length of the polymeric backbone of the molecule. Due to large volume occupancy, such molecules have a better chance of colliding with each other and produce friction resulting in higher viscosity. Thus, an anionic polysaccharide of a given molecular weight shall have a higher viscosity than a linear nonionic polysaccharide of the same molecular weight.

In actual practice it is observed that when carboxymethylation derivatization of a sample of a guar polysaccharide is carried, the resulting carboxymethyl (CM) product, invariably, has a lower viscosity than the starting material. This anomaly is due to the fact that some molecular depolymerization during the derivatization process takes place due to the presence of oxygen (air) and alkali. Also the nonpolymeric byproducts (salts) when left in the final reaction mixture lower the actual amount of CM guar in a given mass of the product.

5.9   GEL FORMATION AND VISCOELASTICITY

Many hydrocolloids are capable of forming gels of different gel strengths, which is dependent on their molecular structure and concentration in addition to other environmental factors in the gel. These factors include total ionic strength, pH of the gel system, and its temperature. Gel-forming hydrocolloids at a certain critical concentration and at a temperature generally higher than room temperature only form a sol. Upon cooling this sol to room temperature or below, it sets into a gel. Such polysaccharides are referred to as *thermogelling* materials, which have a tendency to form a gel at room temperature or lower than that. Such gelling is usually reversible, that is, the gel again melts on heating.

The combined effect of viscosity enhancement and gel-forming behavior is referred as *viscoelasticity*. Viscoelasticity can be examined by determining the effect of an oscillating force on the movement of the gelled material. Some deformation is caused on gelled, viscoelastic hydrocolloid materials by the shear stress. This is referred to as an *elastic deformation*. The contortion caused to the polymer chains into high-energy conformations will again return to the ground energy conformation when the applied force is removed, but the deformation caused by sliding displacement of the chains through the solvent does not return to the ground energy level when the force is removed. Thus, under a constant force the elastic displacement remains the same, but the sliding displacement continues to increase.

5.10   RHEOLOGY MEASURING INSTRUMENTS AND METHODOLOGY

An industry dealing in hydrocolloids (e.g., galactomannans) and the end users of such hydrocolloids generally need an estimate of apparent viscosity of a product at specified concentrations in an aqueous sol and at specified shear rates. This can ensure that a product of required specification is regularly being supplied to the user.

For laboratory testing, a hydrocolloid sol of a known concentration is prepared in distilled water at room or at an elevated temperature. In the latter case, the polysaccharide solution is allowed to cool to room temperature or to a specified temperature before the viscosity measurements are carried out. Stirrers of variable revolutions per minute (rpm) are employed for making the solutions of hydrocolloids.

Viscosity measurement is easily achieved using one of the commercially available and direct reading rotary viscometer or rheometer. For the measurement of apparent viscosities of polymer solutions at concentrations above the critical value (i.e., when the viscosity becomes dependent on shear rate) rotary viscometers are always used.

These instruments are based on measuring the increase in resistance created by the polysaccharide in a sol to a spindle of the instrument rotating at a constant revolution per minute in air or in water, and when it is immersed in the sol fluid and keeps on rotating at the same revolution per minute. The increase in resistance is measured in terms of additional energy required by the instrument to maintain synchronous revolutions per minute. Typical examples of such instruments are the various models of rotary viscometers manufactured by the Brookfield Co., United States.

When the viscosity measurements are to be carried at different temperatures, the vessel containing the hydrocolloid sol can be thermostated. With such commercially available viscometers, the apparent and very high viscosities (~10,000 cps) of rather concentrated polymer sols and pastes can be measured at varying shear rates. The polysaccharide manufacturer supplies all such information, which is needed by the end user. A consumer of hydrocolloids is also supplied with such information as the rate of hydrocolloid hydration, that is, the time required to develop maximum viscosity. Viscosity variation on storing the polysaccharide sol in an incubator at a fixed temperature for over 24 hours is also supplied by the manufacturer. Information about the apparent pH of polysaccharide sol is usually supplied. Particle size distribution of gum powder as determined by sieve analysis is also mentioned in the analysis reports. For hydrocolloids to be used as food additives, a microbiological test report is made for which the necessary laboratory facilities and manpower is required.

REFERENCES

1. Fredrickson, A. G., Principles and Applications of Rheology, Prentice-Hall, Englewood Cliffs, CA, 1981.
2. Lapasin, R. and Pricl, S., Rheology of Industrial Polysaccharides: Theory and Applications, Springer, Berlin, 1999.
3. Glickman, M., Gum Technology in Food Industry, Academic Press, New York, 1969.
4. Glasstone, M. and Lewis, D., Elements of Physical Chemistry, 2nd ed., McMillan & Co. Ltd., London, 1960.

6 Derivatization of Polysaccharides

6.1 INTRODUCTION

Among the industrial polysaccharides, starch and cellulose are homopolysaccharides of glucose, whereas many others are copolysaccharides containing two or more different sugar monomers. Galactomannans are copolysaccharides, and these are composed of mannose and galactose as their sugar monomers. Most of the polysaccharides, which are produced commercially, are generally available in purity, >70%. Under ambient conditions of temperature (25°C –30°C) and humidity (50%–70%), most of the polysaccharides contain 8%–12% equilibrium moisture. Hence, their purity on a moisture-free basis is usually ~90% and for most commercial uses polysaccharides of such purity are satisfactory.[1]

There has been a large difference in the price of various polysaccharide gums, depending upon the abundance of their natural sources, which are plants in most of the cases. More recently a large number of microbial polysaccharide gums are also being produced.[2] Among the galactomannan polysaccharides, those derived from the full-grown, perennial trees, for example, locust bean gum (LBG) from carob tree, which specifically grows in the Mediterranean region of Europe and Africa, and tara gum from pods of tara shrub, which mainly grows in the hilly Andes region of Peru, have limited supply. These gums, which are produced in limited amounts, are much costlier compared to those gums that are derived from annual agriculture plant crops, for example, guar and fenugreek seed crops.

Whereas the starches are metabolized in the human digestive system, most of the other polysaccharides, including the galactomannans, are excreted undigested. However, most of the nonmetabolized polysaccharides are nontoxic to humans and they act as soluble or insoluble dietary fibers. Since the applications of a particular polysaccharide gum are determined by its functional properties rather than its source or its chemical structure, it is sometimes possible to physically modify and chemically derivatize[3] a cheaper and more abundant polysaccharide (e.g., cellulose, starch, and guar gum) into a product that can replace a less abundant and costly gum for specific applications.

With this in view, during the Second World War research was conducted in Germany to derivatize insoluble cellulose into a soluble gum (i.e., carboxymethylcellulose or CMC). CMC was used in Germany as a cheaper substitute for then-scarce gelatin, which is a protein. Most of the early chemistry of polysaccharide derivatization has been developed for cellulose and starch, which are the two most common and abundant polysaccharides. These derivatization methods were later found equally applicable to other polysaccharides, including the galactomannans.

6.2   DIFFERENCE BETWEEN THE WHOLE SEED POWDER AND AN INDUSTRIAL POLYSACCHARIDE[4,5]

For seed polysaccharides, for example, properly isolated maize, potato, or rice starch or guar gum (galactomannan polysaccharide), commercial products are of high purity. These are produced by a good microbiological standard, which are suitable for use as additives in food products. In contrast to this, the whole seed powder, for example wheat flour, which has an edible polysaccharide (starch) as the major component, still contains varying amounts of proteins and many minor constituents (e.g., lipids and various vitamins). Hence, a whole seed powder is termed *seed flour* rather than an industrial-grade polysaccharide. It follows that starting with a specific polysaccharide-bearing seed, certain separation and purification steps are required to get an industrial grade polysaccharide powder. Such methods of galactomannan polysaccharide extractions from many seeds have been developed and these are dealt with in detail in the subsequent chapters on individual galactomannans.

6.3   *NEED FOR MODIFICATION OR DERIVATIZATION OF POLYSACCHARIDES*[3]

An industrial polysaccharide obtained from a plant seed is used in several food and nonfood applications, but in some applications further treatment may be necessary to physically modify it or to chemically derivatize it into a modified or chemically new product. Physically distinct and modified forms of a polysaccharide can differ in their fineness of the powder, or it may be in a deliberately granulated form, produced from a fine powder, or it may be so treated as to make it dispersible in water without lumping or have its viscosity development pattern modified (i.e., slowed or accelerated). Control of the final viscosity of a polysaccharide to a predesired value by blending with other constituents or controlled depolymerization should also be considered as physical modifications because the basic chemical structure of the polydispersed polysaccharide does not change in these operations.

In chemical modification or derivatization, the basic chemical nature and structure of a polysaccharide is changed. As an example, nonionic cellulose can be derivatized into anionic (CMC) or cationic products, which are polyelectrolytes. Similarly the hydrophilic character of a polysaccharide can be reduced or it can be altered into a hydrophobic character by introducing appropriate functional groups into its molecule. For insoluble, native cellulose, derivatization invariably results in enhanced solubility in water or other solvents. A modified or derivatized polysaccharide, as a new product, can have much broader applications when compared to the parent material. The derivatized product can be useful to substitute a high-priced hydrocolloid material. Thus, we can make a rough differentiation between

1. Modified polysaccharides as those in which the basic chemical structure is retained
2. Derivatized product as those in which the basic chemical structure has been changed

6.4 POLYSACCHARIDE DERIVATIVES[6]

Cellulose and starch, being the most abundant and cheap polysaccharides, are frequently derivatized into value-added products.[7] Structural regularity in a linear and unsubstituted polysaccharide, for example, cellulose (β-D-glucan), ivory nut mannan (β-D-mannan), and starch amylose (α-D-glucan) can produce partial crystallinity, which make these insoluble or poorly soluble in water. Structural regularity produces junction zones and crystallinity in polysaccharides to such an extent that in certain cases thermal agitation does not destabilize their junction zones to induce complete water solubility. The result is that their dispersion in water is similar to intertwined spaghetti-like suspension, which has part crystalline and part amorphous regions. These are considered to be poor as hydrocolloids or gum products.

One repeating hexose-sugar monomer unit in a polysaccharide on average contains three hydroxyl groups in addition to the ring and glycoside oxygen as a functional group. These oxygen-carrying functional groups are the centers for hydrogen bonding. When a polysaccharide is placed into water, chances are that it will get fully hydrated, binding clusters of water tenaciously. Water binding action of polysaccharides can be comparable to the dehydrating action of phosphorous pentaoxide. Under such conditions, the polysaccharides should completely dissolve in water and disperse to a molecular level. In actual practice, particularly some linear polysaccharides are prevented from dissolving to a molecular level due to a finite, inter- and intramolecular hydrogen bonding between the strands of linear polymer, resulting in the formation of junction zones and the regions,' crystallinity arising due to structural regularity. This can result in them having poor functional behavior as hydrocolloids or gums.

Cellulose and starch are the cheapest among all the polysaccharides and guar gum is the cheapest among all the galactomannans. Hence, these three have been the most extensively derivatized polysaccharides. Derivatization reactions have also been carried out on other galactomannans, for example, LBG and many others, for which extensive patent literature exist. However, none of the LBG derivatives are now produced commercially.

While useful improvements in properties of the derivatized product is observed in most of these cases, the cost of modification, when added to the already premium price of a gum such as LBG, make such developments less likely to be adopted commercially and remains only as academic patents.

When a modified, or a derivatized, gum is likely to be used in food, the cost of its testing as a safe food additive and the considerable time involved in getting its clearance as a safe food additive must be weighed against expected market, cost and returns from its sale, and risk involved.

6.5 APPROVAL OF DERIVATIZED POLYSACCHARIDES AS FOOD ADDITIVES

Several cellulose and starch derivatives have been approved as additives for food, but so far, none of the derivatized guar and other galactomannan products has been cleared. Derivatives of guar gum have found several applications in nonfood areas. Considering the extensive food safety work[7] carried over a long period on products such as CMC and starch ethers, it may safely be concluded that the analogous guar gum and other galactomannan ethers should also be safe and innocuous as a food additive, but as of yet no manufacturer of such products has applied for clearance by the U.S. Food and Drug Administration or similar agencies in other countries.

Scientists are of the view that the bis-equatorial bond or the β-linkage between the mannose monomers in the mannan backbone of galactomannans, which is similar to that between glucose monomers in cellulose, is not cleaved by human digestive enzymes. The galactomannan derivatives, just like the parent gum, should also be excreted unchanged and should be safe as food additives and act as dietary food fibers. Since the derivatized polysaccharides are less prone to bacterial attack, there is a need for further exploration of these as food additives and hydrocolloids for other industries.

6.6 PRODUCTION OF GALACTOMANNAN DERIVATIVES[8]

Nearly all the methods for commercial derivatization of polysaccharides, originally developed for cellulose, have also been applied to guar gum and patented. Patents on similar lines have also been reported for other galactomannans, particularly for LBG, but commercial production of LBG derivatives is currently a minor activity. Hence, we will discuss galactomannan derivatives mainly with respect to guar gum.

Production of guar gum derivatives was initially carried out using a slurry of the gum powder in water and organic solvents mixture, for example, lower alcohols (methanol, ethanol, or 2-propanol) or acetone. In a typical case of preparation of carboxymethylated guar gum, slurry of the gum powder is reacted with monochloroacetic acid (MCA) or its sodium salt (SMCA) in the presence of an alkali (sodium hydroxide). In this reaction, an alkali acts as a catalyst besides being a reactant. The reaction mixture is stirred at 60°C–80°C and the reaction is complete in 1 to 2 hours, after which the product derivative is filtered, washed with the same solvent mixture, and dried. The solvents are recovered for reuse. Polymer degradation by alkali and oxygen, and consequently lowering of viscosity, is minimized by preventing an air exposer of the reaction mixture and nitrogen blanketing, and avoiding a large excess of alkali. A derivatized product prepared by the slurry method is purer, being free of by-products and unreacted reagents.

Slurry methods are now only of academic interest and most of the current commercial methods are based on the use of gum powder or split. Usual precautions to minimize air and alkali degradation are followed to maintain high viscosity of the product derivative made from gum powder or split.

6.7 REACTIVITY OF VARIOUS HYDROXYL GROUPS OF A POLYSACCHARIDE DURING A DERIVATIZATION REACTION

During an early structural study of polysaccharides (e.g., cellulose and starch), their derivatization chemistry (e.g., methylation) was developed as a laboratory tool and it was much later that these reactions were adopted on an industrial scale.

A hexose unit in a β,1→4-linked or bis-equatorially linked polysaccharide molecule such as cellulose has three free hydroxyl groups on numbers 2, 3, and 6 carbon atoms of the pyranose ring. These hydroxyl groups are accessible to derivatization such as etherification and esterification reactions. During substitution in a derivatization reaction, the position taken by a substituent group depends upon several factors, including the accessibility of the derivatizing reagent to the hydroxyl groups, which are sometimes buried into the crystalline region of the polysaccharide and hence can escape derivatization. Other hydroxyl groups, which are better exposed for reaction in the amorphous regions of the polysaccharide, react better.

Other factors determining the reactivity of various hydroxyl groups include inductive effect, steric factor, and the conformation of polysaccharides. For starch and cellulose, the reactivity order of the various hydroxyl groups in a glucose unit has been found to be in the order C-2 > C-6 > C-3. For steric consideration, C-6 primary hydroxyl appears to be less hindered, but inductive effect, due to glycoside oxygen, increases the reactivity of C2-OH. Not much investigation has been carried out to determine the reactivity of these (2, 3, and 6) hydroxyl groups in galactomannans, but the reactivity order seems to be in the same sequence as that of a glucose monomer in cellulose or starch. In the case of galactomannans, substitution can occur in mannose as well as in the galactose unit of the polysaccharide.

6.8 DEGREE OF SUBSTITUTION AND MOLECULAR SUBSTITUTION[9]

As mentioned earlier, there are on average three hydroxyl groups on a hexopyranose sugar monomer ($C_6H_{10}O_5$) unit in a polysaccharide, and the equivalent weight of this repeat unit in a polysaccharide is 162. Three is the maximum number of ether (-OR) or ester groups that can be introduced in one sugar monomer unit. The average number of substituents that are actually present in a sugar monomer of a polysaccharide derivative is called the *degree of substitution* (DS) and it can have a maximum value of three when the polysaccharide is fully substituted. In actual commercial derivatization practice, the DS is kept low enough to serve the purpose of modification, for example, to improve the solubility, and it seldom exceeds one. Thus, in a CMC of DS equals 0.1, only one hydroxyl group out of ten glucopyranose monomers carries a substituent, that is, the carboxymethyl (CM) ether group. When DS equals 1.0, on average there should be one hydroxyl in each glucopyranose etherified. Generally a DS of ~0.1 is enough to bring about a dramatic change in the solubility of a polysaccharide.

In derivatization reactions involving hydroxyalkylation, using the alkene oxide reagents (e.g., by hydroxyethylation or hydroxypropylation), the groups introduced are -OCH$_2$CH$_2$OH (hydroxyethyl) and -OCH$_2$CH(CH$_3$)OH (hydroxypropyl). These substituent groups also have a hydroxyl group, which is even less hindered than pyranose ring hydroxyls. In the subsequent reaction, there is a possibility of another alkene oxide molecule reacting with this side chain hydroxyl. When this happens, a sort of graft of polyethylene oxide-type side chain [-O(CH$_2$CH$_2$O)$_n$CH$_2$CH$_2$OH] can be formed and substitution of more than one alkene oxide on the same hydroxyl group takes place. In such case the average number of groups introduced per hydroxyl is called *molecular substitution* or MS.

6.9 WHY WAS DERIVATIZATION TECHNOLOGY DEVELOPED?[10]

One of the oldest used, edible, nonpolysaccharide hydrocolloids or a gum additive in food products has been the protein gelatin. Gelatin is derived from bones and hides of animals, and it has been extensively used as a thickener and gelling agent in the Western countries. Due to its scarcity and high cost, the substitute CMC, which is derived from cellulose, was developed in Germany. This has been the starting point for the industrial technology for the production of polysaccharide derivatives. Among the cheapest and most abundant polysaccharides, native cellulose has the limitation of being water insoluble and hence it cannot be used as a food hydrocolloid or gum. It can be converted into a water-soluble carboxymethyl derivative or CMC, which is now extensively used as a food hydrocolloid.

Linear polysaccharide molecules have a strong tendency to associate and consequently they have poor solubility in water. The presence of a large number of short grafts on a linear polysaccharide reduces such an association and induces water solubility. The extent to which a linear polysaccharide can be made water soluble depends on the frequency of the grafts and the mode of their placement along the linear polymer backbone. Additionally, if the side-chain grafts carry a charge, the solubility is further improved due to separation of molecular chains by columbic repulsion, as in the case of CMC.

It follows that when an insoluble, linear, homoglycan (e.g., cellulose) is derivatized by introducing side-chain groups (e.g., carboxymethyl ether) at regular intervals on the backbone, then the polymer strands are prevented from coming close to associate, resulting in water solubility. The net result is conversion of insoluble cellulose into a soluble gum (CMC) or a hydrocolloid. Surprisingly, even DS as low as 0.01 to 0.04 has been found to be effective for solubilization of cellulose into an entirely new product with newer applications.

6.10 DERIVATIZATION USING GUM POWDER (CARBOXYMETHYLATION)

When guar gum powder is used for carboxymethylation, concentrated solutions of the reagents SMCA and alkali are sprayed on well-mixed powder in a suitable, horizontal blender-reactor, preventing lump formation. With a good spraying and mixing arrangement, up to 50% moisture can be present in the powder and still it should be in a well-mixing and nonlumping powder form. Solid SMCA can also be well mixed with dry gum powder (moisture 10%–15%), followed by spraying the requisite amount of 20%–50% alkali, with thorough mixing and the temperature being maintained at 60°C–80°C, for 1 to 2 hours. Salt produced as a by-product and hydrolyzed reagent are permitted to remain in the product if it is meant for technical use (e.g., textile printing paste thickener).

The reaction of carboxymethylation of a polysaccharide can be represented as follows

$$PS\text{-}OH + NaOH \rightarrow PS\text{-}ONa + H_2O$$
$$PS\text{-}ONa + ClCH_2COONa \rightarrow PS\text{-}O\text{-}CH_2COONa + NaCl$$

where PS-OH = polysaccharide.

In a printing paste thickener, CM guar gum works better by preventing chocking of a printing screen and it is easy to remove after the washing of a printed cloth.

6.11   DERIVATIZATION USING GUAR SPLIT[8]

When guar split is used for derivatization, the hydrated split particles, containing up to 50% weight/volume (W/v) water, function like a gel, which being permeable to low molecular weight reagents (i.e., SMCA and alkali) react well as the gum powder or its slurry. In the gum splits, the polysaccharide is naturally encased in a cellulose cell wall, which permits the diffusion of the reagents inside, and the reaction takes place well inside the cell wall, with the same ease and efficiency as with powder or slurry.

The cellular structure of the split remains undisrupted during derivatization when the temperature is kept in the range 60°C–80°C. Since the cellular structure is not disrupted during the derivatization reaction, the derivatized splits can be washed if so desired, preferably after a surface cross-linking with borax, to remove the by-products (i.e., NaCl, unreacted SMCA, NaOH, and hydrolyzed reactant, which is glycolic acid). This is followed by partial drying, flaking, and powdering of derivatized splits.

6.12   POLYSACCHARIDE DERIVATIVES CURRENTLY BEING MANUFACTURED AND PATENTED

It was mentioned that the cost of a polysaccharide derivative is determined from the cost of the original raw material (polysaccharide) and the cost incurred in the derivatization process and chemicals used. The following derivatives have been reported for guar and numerous other polysaccharides in patents.

Nonionic derivatives—Alkyl (methyl, ethyl, and allyl) and hydroxyalkyl (hydroxyethyl or HE and hydroxypropyl or HP)

Anionic derivatives—Carboxymethyl (CM) and carboxymethyl hydroxypropyl (CMHP; a double derivative), carboxyethyl, sulfate esters, phosphate esters, and sulfonic (sulfoalkyl)

Cationic derivatives—Pri-, sec-, and tert-aminoalkyl and quaternary derivatives

Commonly available derivatization reagents for polysaccharides are the lower alkyl halides and lower alkene oxides, monochloroacetic acid or its salt, and a variety of substituted halohydrins and epoxides. The mechanism, which is followed in the etherification reactions involving alkyl halides and monochloroacetic is the classical

Williamson's ether synthesis reaction. In the reactions involving alkene oxides and substituted alkene oxides, nucleophilic addition of a hydroxyl group to the epoxide ring system takes place. These etherification reagents are used along with a strong base (e.g., NaOH), which, besides acting as a catalyst, also produces an active epoxide from certain chlorohydrin reagents.

In Table 6.1, some details about the carbohydrate derivatives, derivatizing reagents, and their molecular weights are given. For reagents 2, 3, 5, and 6, the reaction takes place via addition of an epoxide group. An epoxide group is formed with reagents 5 and 6 from a chlorohydrin reagent during the reaction by action of alkali. This is followed by the addition of a polysaccharide OH to the epoxide.

The following general reaction scheme represents the formation of an active epoxide carrying a cationic or anionic functional group from these chlorohydrin reagents.

$$\text{Cl-CH}_2\text{-CHOH-X} + \text{NaOH} = \text{CH}_2\text{-CH}_2\text{-X} + \text{NaCl}$$
$$\overset{\diagdown\ \diagup}{\underset{O}{}}$$

(Formation of an epoxide from a chlorohydrin)

$$\text{PS-OH} + \text{CH}_2\text{-CH}_2\text{-X} = \text{PS-O-CH}_2(\text{OH})\text{-CH}_2\text{-X}$$
$$\overset{\diagdown\ \diagup}{\underset{O}{}}$$

(Addition of epoxide group to a polysaccharide-OH)

where PS = the polysaccharide and the group, and X = H, CH_3, $CH_2\,N^+(CH_3)_3\,Cl^-$, and $CH_2\,SO_3Na$ for derivatives, 2, 3, 5, and 6, respectively (Table 6.1).

In most of the derivatizing processes, it is an advantage to know the mole ratio of polysaccharide monomer (equivalent weight of hexose unit, i.e., $C_6H_{10}O_5 = 162$) and the reagent used for derivatizing. The molecular weights of the common derivatizing reagents are mentioned in Table 6.1.

TABLE 6.1
Some Important Polysaccharide Derivatives

Derivative (-OR)	Nature of Group R	Reagent (Mol. Wt.)
Methyl (-OCH$_3$)	Hydrophobic	Methyl chloride (50.5)
Hydroxyethyl (-OCH$_2$-CH$_2$OH)	Hydrophobic	Ethylene oxide (44)
Hydroxypropyl (-OCH$_2$-CHOH-CH$_3$)	Hydrophilic	Propylene oxide (58)
Carboxymethyl (-OCH$_2$-COOH)	Weakly anionic	MCA or SMCA (94.5, 116.5)
Sulfonic salt (OCH$_2$-CH(OH) CH$_2$SO$_2$OH)	Strongly anionic	Sodium (3-chloro-2-hydroxypropane) sulfonate (196.5)
Quaternary ammonium	Strongly cationic	(3-chloro-2-hydroxypropyl)-trimethyl-ammonium chloride (188)

Generally the yield in such etherification reactions is >70%. The methods for determining DS for different substituents will be discussed later, but a satisfactory method for determining MS has not been developed.

6.13   COMMON DERIVATIZING REAGENTS AND REACTIONS[11]

Alkyl chlorides (CH_3Cl, C_2H_5Cl) are invariably used for making nonionic alkyl derivatives. Lower alkene oxides, typically ethylene oxide and propylene oxide, are used for making HE and HP derivatives, respectively. Lower alkene oxides and alkyl chlorides being gaseous reactants require a temperature controlled, jacketed pressure reactor, with gas and liquid reagent feeding and metering arrangements.

MCA or SMCA is used for making CM derivatives. Equations representing the addition of acrylonitrile or acrylamide to the hydroxyl group followed by hydrolysis have been used for making carboxyethyl derivative as shown next:

$$PS-OH + CH_2=CH_2CN \rightarrow PS-O-CH_2-CH_2CN,$$
$$PS-O-CH_2-CH_2CN + 2H_2O \rightarrow PS-O-CH_2-CH_2COOH + NH_3$$

Sulfate esters are made from anhydrous polysaccharide, using the sulfurtrioxide-tertiary amine (pyridine) complex, while the reaction of dibasic sodium phosphate is used for making phosphate esters. 3-Chloro-2-hydroxypropyl-trimethyl-ammonium chloride has been used for making quaternary-group substituted polysaccharides.

Chemical methods for determination of DS in various polysaccharide derivatives have been developed. These methods are based on particular functional group determination methods. Most of these methods are described in organic analytical chemistry books. Use of nuclear magnetic resonance (NMR) spectroscopic methods, such as those based on specific proton counting, for example, that of methyl proton in hydroxypropyl-CH_2-$CHOH$-CH_3 derivative, can also be done.

6.14   DETERMINATION OF DEGREE OF SUBSTITUTION[10]

It has been found that the reaction efficiency in most of the etherification reaction is quite high, in the range of 70%–75%. Hence, as an approximation, an estimate of the DS in a particular derivatization reaction can be made from the mole ratio of the polysaccharide monomer ($C_6H_{10}O_5 = 162$) and that of the derivatization reagent (Table 6.1) used, considering the reaction efficiency to be ~70%–80%. In most of the patents on derivatization of polysaccharides, no mention is made of the mole ratios of the reagent to the polysaccharide monomer employed, and only a weight ratio of polysaccharide, reagent, and alkali is mentioned.

The following mathematical relations are useful for calculating the DS. When n is the number of sugar monomer units ($C_6H_{10}O_5 = 162$) in a fragment of polysaccharide, which contains one substituent (OR), for example, the ether group, introduced during derivatization (etherification) of -OH, the DS = 1/n and this polysaccharide fragment can be represented as

$$(C_6H_{10}O_5)n-1.C_6H_9O_4.OR.$$

The molecular weight of this fragment equals $(n - 1)162 + (161 + r)$, where r is the equivalent weight of group R. Methods for the determination of DS can now be based on the determination of functional group R in such a fragment.

The following is a list of the various functional groups R in different derivatives (Table 6.1) are

1. Methyl = ($-CH_3$),
2. Hydroxyethyl = ($-CH_2-CH_2OH$)
3. Hydroxypropyl = ($-CH_2-CHOH-CH_3$),
4. Carboxymethyl (free acid, salt) or CM = ($-CH_2COOH$, $-CH_2COO\,Na$)
5. Quaternary or 2-hydroxypropyl-trimethyl-ammonium chloride = ($-CH_2-CHOH-CH_2\,N^+(CH_3)_3Cl^-$)
6. Sulfonate (salt) or 2-hydroxypropanesulfonate = ($-CH_2-CHOH-CH_2\,SO_3Na$)

6.15   TYPICAL METHODS FOR DETERMINATION OF DS

As a typical example, the determination of DS in a CM derivative will be described, which involves

1. Preparation of an analytically pure sample for the determination of DS—Commercial samples of polysaccharide derivatives may contain by-products (e.g., salts, alkalis, and hydrolyzed reagent). These are frequently and deliberately left in the nonfood, technical products. Hence, it is necessary to make a pure (analytical) sample of the derivative for its analysis and for determining the DS.

   A 5–10 g derivative sample is dissolved in a minimum amount of hot distilled water and any insoluble residue is removed by filtration/centrifugation. The clarified filtrate is then pored into twice its volume of alcohol (ethanol or 2-propanol) to precipitate the pure salt of CM polysaccharide. The precipitate is then filtered, washed with alcohol, and dried at 105°C. This general procedure can also be used to prepare analytically pure samples of other polysaccharide derivatives.

2. Determination of carboxylic group in the sample—A method based on a nonaqueous titration of the polysaccharide carboxylic-salt by titration against standard perchloric acid solution (in glacial acetic acid) as a titrant and gentian violet as an indicator can be used in this case.

   A more frequently used gravimetric method is based on the combustion of a weighed amount of an analytical sample of the derivative (CM salt) in a preweighed crucible when the quantitative formation of Na_2CO_3 results. This can be weighed or determined by titration against standard acid.

3. Calculation—Suppose, W gram of CM-derivative sample produce w gram of Na_2CO_3 on combustion, then the molecular weight of a fragment of the derivative (discussed earlier) having one CH_2COONa group per "n" sugar-units shall produce half mole (53 part) of Na_2CO_3. Hence, the molecular

weight of the fragment, $(C_6H_{10}O_5)n\text{-}1\text{-}C_6H_9O_4\text{-}O\text{-}CH_2COONa = (W \times 53)/w$. From this relation, the DS (1/n) can be computed. For determination of other functional groups the following methods can be used.

4. Alkyl and hydroxyalkyl—A modified Zeisal's method has been used for alkoxy (methoxyl) group determination. For the hydroxypropyl group, a Kuhn-Roth method for C-methyl group determination has been used by the author. This is based on quantitative oxidation of the HP derivative of a polysaccharide by chromic acid into acetic acid. Acetic acid thus produced is steam-distilled and determined by titration against the standard base. NMR determination of methyl proton in a sample can also be followed.

5. Quaternary ammonium halide—In this case, Kjeldahl's method of nitrogen determination is not considered suitable because of the possible presence of a protein impurity in the sample. A more accurate method is based on the determination of ionic chloride in a weighed sample by silver nitrate titration (Mohr's method).

6. Sulfate/sulfonate group—In such derivatives, a gravimetric method based on combustion of a weighed amount of a sample derivative to quantitatively produce Na_2SO_4 has been followed by the author, which is exactly similar to the determination of CM salt.

Overall it can be seen that a chemist having experience of organic analytical methods can devise other similar methods for any other functional groups in polysaccharide derivatives.

6.16 PREPARATION OF COMMON DERIVATIZING REAGENTS

Derivatization of a polysaccharide into a sulfonic group–containing product, which was developed by the author has been referred to earlier. For this, a sulfonic acid group containing epoxide reagent was synthesized and loaded onto a polysaccharide, according to the following reaction scheme.

$$Cl\ CH_2. CH. CH_2 + NaHSO_3 \longrightarrow Cl\ CH_2. CH(OH). CH_2.SO_3.Na \quad \text{--- (a)}$$
$$\underset{O}{\backslash\ /}$$

$$Cl\ CH_2. CH(OH). CH_2SO_3Na + NaOH \longrightarrow CH_2. CH. CH_2SO_3Na + NaCl + H_2O \quad \text{--- (b)}$$
$$\underset{O}{\backslash\ /}$$

$$PS\text{-}OH + CH_2. CH. CH_2SO_3Na \longrightarrow PS\text{-}O\text{-}CH_2. CH(OH). CH_2SO_3Na \quad \text{--- (c)}$$
$$\underset{O}{\backslash\ /}$$

Part (a) is the synthesis of chlorohydrin reagent containing sulfonate group, (b) is an in situ formation of the reactive epoxide-form of the reagent from the chlorohydrin, and (c) is loading of the epoxide reagent onto the polysaccharide molecule.

The reaction scheme involves the nucleophilic addition of bisulfite ion to the epoxide group in epichlorohydrin. In a similar way, several other nucleophiles (e.g., cyanide, sulfide, nitro, etc.) can also be added to the epoxide group in epichlorohydrin to synthesize epoxide reagents containing specified functional groups. These reagents can then be reacted with polysaccharides in the presence of an alkali to give derivatives containing several different functional groups. This amounts to a very general method for making several derivatized polysaccharides, which can have some specific applications. In one particular application of the thiol (SH) group containing, and cross-linked, insoluble-bead form of guar gum, which is made insoluble by cross-linking with epiclorohydrin, were prepared by the author. These beads have been used in the removal of traces of heavy metals (Pb, Hg, Ag, etc.) from water. When the bead form of this reagent is added to water, the thiol group binds the traces of heavy metals, and these can be removed by filtering off the beads.

REFERENCES

1. Aspinall, G. O. Polysaccharides, Oxford University Press, Oxford, 1970.
2. Glickman, M., Gum Technology in Food Industry, Academic Press, New York, 1969.
3. Whistler, R. L. and BeMiller, J. N., eds., Industrial Gums, 3rd ed., Academic Press, New York, 1993, 181.
4. Paroda, R. S. and Arora, S. K., Guar, Its Improvement and Management, The Indian Society of Forage Research, Hissar, India.
5. Anonymous, A Study on Guar Gum, Agricultural and Processed Food Products Export Development Authority, New Delhi. Dalal Consultants and Engineers, Limited, Ahmedabad, India.
6. Radley, J. A., Ed., Starch and Its Derivatives, Vol. 1, 3rd ed., Chapman & Hall, London, 1953.
7. Code of Federal Regulations, Title 21, No. 19530, and similar others, U.S. Govt. Printing Office, Washington D.C.
8. Nordgren, R., Jones, D. A., and Wittcoff, H. A., U.S. Patent 3,723,408, 1973, Assigned to General Mills Chemicals Inc.
9. McLaughlin, R. R. and Herbest, J. H. E., Can. J. Res., 28B, 1950, 737.
10. Purves, C. B., Cellulose and cellulose derivatives, In High Polymer, Vol. 5, 2nd ed., Interscience, New York, 882–938.
11. Mathur, N. K., Unpublished. Based on manufacturing practice followed at several guar gum industries at Jodhpur.

7 Guar Gum
The Premium Galactomannan

7.1 INTRODUCTION AND TRANSDOMESTICATION OF GUAR CROP

It has been suggested that guar is not an indigenous crop of the Indian subcontinent. The exact history of transdomestication of the guar crop into the Indian subcontinent is also uncertain. As suggested by Hymowitz,[1] the guar crop probably evolved from its wild predecessor in the Middle East, from where it was brought into India as fodder by the Arab horse-traders in their merchant ships. Even this suggestion has now found some contradictions.

Once agriculture of the guar crop was started in India, the semiarid climate of Northwestern India (Rajasthan state and some nearby areas in the Haryana, Punjab, and Gujrat states) was found very suitable for growth. Over many centuries of its agricultural cultivation and crop selection in India, it has probably resulted into the present day guar crop. A detailed history of the guar crop and its spread into other continents has been discussed in Whistler's monograph on industrial gums, and Whistler and Hymowitz's monograph on guar.[2]

Even after cultivation for many centuries as an agriculture crop, guar remained only as a minor crop of a remotely underdeveloped area of the semiarid region area of Northwestern India. It was during the mid-20th century and only due to a much later realization that guar seed could be a source of a unique galactomannan polysaccharide (gum) that brought it to the current status as a useful industrial crop.

7.2 GUAR SEED AS A SOURCE OF AN INDUSTRIAL GALACTOMANNAN GUM[3]

Most of the cereal grains and many edible legume seeds have starches as their reserve polysaccharides, but some legume seeds contain galactomannans as reserve polysaccharides.[4,5] Thus, the seeds of many annual legume crops (guar and fenugreek) as well as full-grown legume trees and shrubs (carob and tara) have been sources of industrial galactomannans. Several food and nonfood industries use many different galactomannans as industrial hydrocolloids. Legume-based foods, which are popular in Asian countries, owe their characteristic, taste, and flavor to the presence of many minor polysaccharides, along with starches in their seeds.[6]

Industrial galactomannans in current production have been derived from the seeds of annual agricultural crops (e.g., guar and fenugreek), seeds of certain wild, annual herbs (e.g., *Cassia tora* and *Sesbania bispinosa*) as well as from full-grown perennial trees and shrubs (e.g., carob [also known as the locust tree] and tara shrub).[7]

Any germinating plant seed embryo shall use its reserve polysaccharide as the source of carbon in its initial phase of growth. It is only after the plant has emerged out of the soil that it starts doing photosynthesis, and uses soil water and atmospheric carbon dioxide as the source of carbon for its growth.

An extensive survey of seed polysaccharides from legume plants has revealed that many legume seeds contain galactomannans as their endospermic mucilage.[8] Some of these seed gums have been known and used since ancient times. Thus, the use of the pods of the carob tree was known even during the Biblical period and carob seed powder finds a mention in the Bible as a constituent of St. John's bread. Among the various legume seed galactomannans now known, the one derived from the locust bean,[3] or the pod of the carob tree (*Cerartonia siliqua*), is native to the Mediterranean region of Europe and Africa. The carob tree has been one of the oldest sources of an industrial hydrocolloid, known as locust bean gum or LBG.

Indian flora is particularly rich in legume plants. Kapoor and co-workers, at the National Botanical Research Institute (CSIR, India), have made an extensive survey of galactomannan-bearing legume plants of India and Southeast Asia.[9] An important galactomannan-bearing legume plant widely cultivated in India is guar (*Cyamopsis tetragonolobus*). Guar is an annual crop that has been cultivated for centuries on the Indian subcontinent (India and Pakistan).[6,7]

Guar seeds were principally used as a feed for cattle and its unripened green pod was, to a much lesser extent, used as a vegetable in human food. It was much later during the years 1944 to 1953 that guar seed was investigated and commercialized as a source of an industrial galactomannan. The endosperm polysaccharide from guar seed (i.e., guar gum) has also been called *guaran*.[8,10] According to the chemical nomenclature rules, it is called a *galactomannan polysaccharide* (a copolymer of mannose and galactose sugars). It is extracted from the seed of the guar plant, where it serves as a reserve polysaccharide for the embryonic plant and for water storage during the period of water stress.

7.3   DEVELOPMENT OF GUAR CROP AS THE SOURCE OF AN INDUSTRIAL GALACTOMANNAN[2]

The galactomannan gum from guar seed was developed as an industrial commodity as an exigency of World War II. Prior to this, LBG was being used to enhance the recovery of cellulose pulp for paper making and as a strength additive to paper pulp by the American paper industry. When the supply of LBG got severely restricted during World War II due to the occupation of most LBG-producing Mediterranean regions by Germany and Italy, the Institute of Paper Chemistry in Appleton, Wisconsin was looking for a suitable substitute for LBG. It was then that guar gum was found to be a suitable, alternative additive to paper pulp.[10]

Once guar crop was established as a source of an industrial gum, the demand for guar gum in the United States, Japan, and Europe rapidly increased. Attempts were then made to cultivate guar crop in the semiarid regions of Texas and Arizona in the United States and similar regions of Australia and Africa. Such attempts, however, did not prove very successful, partly due to economic consideration. Currently India and Pakistan combined, continue to be the major agricultural producer (>90%) of guar crop and processor of guar seed into an industrial polysaccharide gum.

General Mills Inc. and Stein, Hall & Co. in the United States pioneered milling technology to extract out guar gum from its seed. Development of guar seed endosperm powder manufacturing technology has made the use of guar gum for industrial applications possible. Once guar gum was introduced in the United States in the paper industry to substitute for LBG, its applications in several other fields, including oil-well drilling, textile printing, slurry explosives, and as a food additive, rapidly increased.[11] During the past four decades, guar gum consumption in the United States (Table 7.1) and elsewhere in the world got a further boost after introduction of its anionic, cationic, hydrophobic anionic (double derivative), and nonionic (hydroxyalkyl) derivatives. These guar gum derivatives have found numerous industrial applications. Current trends of guar gum consumption indicate that more such intensive development efforts are likely to continue in years ahead.[10]

Commercial guar gum of food as well as of many technical grades and its derivatives are now sold around the world under various trade names, including SuperCol from General Mills and Jaguar from Rhodia/Stein, Hall & Co., and their European subsidiaries. Some European companies, including Mey Hall Co. (product name Meypro gum) of Switzerland and Cesalpinia S.p.A. (product name Dealca) of Italy also manufacture guar gum.[11]

However, it may be noted that with current trends of globalization and outsourcing of production, most of the guar gum powder, including its derivatives, are now being manufactured in India and Pakistan for these multinational companies and exported all over the world. Indian guar-gum-producing companies also manufacture and sell guar gum products under their own trade names, for example, Shree Ram in Jodhpur manufactures Ramcol; Sunita Minechem Industries in Jodhpur manufactures Sun-Gum; Hindustan Gum in Hissar manufactures HiGum; and Lucid Colloids (formerly a part of Indian Gum Industry or IGI) in Jodhpur produces Edicol brand food-grade guar gum.

TABLE 7.1
Yearwise Estimate of Guar Gum
Consumption in United States

Year	Consumption (lbs)
1954	2,500,000
1955	5,500,000
1956	12,000,000
1957	15,000,000
1960	20,000,000
1965	25,000,000
1980	40,000,000
1990	65,000,000
2005	100,000,000

The city of Jodhpur in the northwestern part of the Rajasthan state in India is the major guar gum (>50% of world's total production) processing center. Many guar gum manufacturing industrial units are located in various other Indian cities, for example, Hindustan Gum (Hissar city in Haryana state), H. B. Gum Industries and Indian Gum Industries (Ahamdabad city in Gujrat state), and Vikas Gum Industry (Sriganganagar city in Rajasthan).

7.4 ESTIMATES OF GUAR GUM PRODUCTION AND ITS CONSUMPTION[11]

It was mentioned earlier that besides guar, there are many other legume plant seeds bearing galactomannan polysaccharides. Only a few of these are currently being used for commercial production of seed gums. For economical considerations, the preferred galactomannans under commercial production are derived from annual crops rather than perennial trees. By putting more land under the cultivation of annual crops (e.g., guar), demands can be met when there is an increase in the world market demand.

Production of guar gum during any specific year depends on the availability of guar seed, which in turn depends on its annual crop yield.[2] Guar crop sowing in India varies from year to year and depends on monsoon season in the Indian subcontinent. During any specific year, by putting additional requisite amounts of land under its cultivation more guar can be produced. In contrast to this, for a perennial tree, such as carob, the fruit bearing occurs only after the tree has matured 10 years and then the tree continues to bear fruit for over a span of nearly 60 years. However, the land occupied in tree plantation is permanent and it cannot be alternated for growing other crops.

In the early 1970s, the total guar seed production was about 500,000 tons, which nearly doubled in 1990s. Yet, there have been large variations in annual guar seed production. For example, due to unfavorable monsoon conditions and crop failure in 1987–1988, the seed production was only 100,000 tons; during 1990–1991, it was more than ten times (1 million tons) this amount.

As mentioned, earlier guar has been mainly a rain-fed crop in India and dependent on the southwest monsoon. From 1995 onward guar was also being grown in irrigated areas of Rajasthan state and its neighboring states (Gujarat, Punjab, and Haryana) in India, and the neighboring Sindh province in Pakistan. This has resulted in less variation in the annual guar seed production and consequently a reduction in large market-price variation of guar gum.

Annual world production of guar gum is nearly 130,000 tons. Of this about 90% (117,000 tons) is produced in India and Pakistan. Indian production of guar gum amounts to about 70% (82,000 tons) and the remaining 30% of it comes from Pakistan. Nearly 50% of Indian production is done in Jodhpur. Total contribution of the state of Rajasthan towards guar gum production in India is nearly 80%.

Dalal Consultants and Engineers, at the request of the Agricultural and Processed Food Products Export Development Authority (APEDA), Government of India, completed an extensive survey of the world's guar gum production.[11] Similar information was earlier compiled in Whistler and Hymowitz's monogram on guar.[2]

The consumption of guar gum and its derivatives in the United States amounted to about 60,000 Mt in the mid-1970s, reaching nearly 90,000 Mt in 1999. Guar powder as well as guar split (called *dal* in Hindi) is exported from India and Pakistan to Western countries, the amounts of which varies. When the export of guar split increases, it is an indication that the worldwide consumption of modified and value-added guar products (derivatives and modified guar gum products), which are manufactured outside India, has increased. Previously, Indian guar manufacturers and exporters were largely content with the export of plain (unmodified) guar gum powder or the split rather than its modified products. During the past two decades, attempts at making specialty guar products in India have been made. Some larger and progressive Indian industries (Lucid Colloids and Hindustan Gum) are now manufacturing some value-added guar products. Certain incentives and facilities by the Indian government to guar gum manufacturers and exporters have been occasionally provided.

7.5 GENERAL STRUCTURAL FEATURES OF GUAR GALACTOMANNAN[10,12]

Seed galactomannans from different legume plants have some common structural features, though they differ considerably in their molecular weight, ratio of the component sugars (mannose-to-galactose ratio, or M:G), the mode of placement of single galactose stubs on the linear mannose backbone in their molecule, and their functional properties. These structural variations have resulted in the availability of a very broad spectrum of these galactomannan-based gums, and provided a wide range of applications based on their well-defined structures and functionalities.

Guar polysaccharide has the common structural features of most other legume seed galactomannans, that is, it has a linear, polymer backbone of $\beta(1{\rightarrow}4)$-linked mannopyranose repeat units, which are randomly substituted by single, $\alpha(1{\rightarrow}6)$-linked galactose grafts or side chains.

Different plant seed galactomannans and their water-solubility-based fractions are known to have variable M:G ratios and different molecular weight ranges. Thus, fenugreek seed polysaccharide has the highest percentage (~48%) of galactose and nearly all its backbone mannose units are substituted by single galactose grafts, making the M:G ratio close to 1:1. Guar gum (M:G = 1.85:1) has the second highest percentage of galactose (>33%), and LBG (M:G – 4:1) has an even lower (~25%) number of galactose grafts. Viscosity, water solubility, and other functional properties, which determine the applications of any galactomannan, depends on the molecular weight and the M:G ratio as well as the distribution of galactose grafts along the polymer backbone.

7.6 GUAR GUM OR GUARAN: A GALACTOMANNAN POLYSACCHARIDE[10]

The LBG-producing carob tree in the Mediterranean region takes about 10 years to mature before it starts bearing pods. It has a life span of more than 60 years. In contrast to this, the guar plant grows as an annual crop, and it takes only 3 to 5 months for the crop to mature. For this and certain other reasons, only a few of these annual legume plants, typically guar and fenugreek, have been utilized for commercial extraction of galactomannan gum from their seeds. Being the most abundant and economical gum, guar galactomannan finds maximum applications as a plant hydrocolloid in food and nonfood industries. It has become the consumer's first choice in any application. Among the industrial galactomannans, guar gum occupies the topmost position, in the quantity produced. Applications to which guar gum are put and its total market value are highest among such hydrocolloids.

As mentioned earlier guar gum was basically developed only as a substitute for LBG, in the U.S. paper industry during and just after World War II. It was not until 1953 that the commercial production of guar gum was started in the United States. At that time the whole guar seed was imported into the United States from India and Pakistan. Initially it was the two American companies, namely, General Mills and Stein, Hall & Co., that developed the technology for separation of guar endosperm or the guar-split from the seed, and grinding it to a fine mesh powder. Later these companies were merged with larger multinational companies, manufacturing a variety of water-soluble gums or hydrocolloids. Food hydrocolloid divisions of these companies are continuing the production and marketing of guar gum along with other food hydrocolloids. The original trade names SuperCol (General Mills) and Jaguar (Stein, Hall & Co.) for guar gum are still being used. These companies also have their European counterparts. There are also certain independent manufacturers of guar gum products in Europe, for example, Cesalpinia of Italy and Mey Hall of Switzerland.

7.7 DEVELOPMENT OF GUAR SEED PROCESSING TECHNOLOGY IN INDIA[11]

It was earlier mentioned that India and Pakistan are the two major countries in the world that produce and process guar seed. Initially during the mid-1950s, American guar gum producing companies imported whole guar seed for processing into gum powder. Since this resulted in a shortage of protein-rich guar germ being used as an animal feed in India, the Indian government imposed restrictions on the export of whole guar seed. In a way, this restriction forced American companies to collaborate with some selected companies in India and Pakistan and transfer gum-producing technology to these companies.

Guar processing technology was initially given by U.S. companies to certain selected industries (e.g., the Hindustan Gum Co., the Indian Gum Industry, H.B. Gum, and Lucid Colloids [formerly Indian Gum Industry]). For a limited period these companies worked in collaboration with American companies, which also helped them in export of the gum. Guar gum manufacturing technology is now well adopted in India and Pakistan, with many improvements over the original U.S. technology. A large number of guar gum manufacturing companies, in India are ISO certified, follow good manufacturing practices (GMP), and do critical control of their process and the products. This has made the guar gum produced in India an internationally accepted commodity.

Currently the guar gum sold to the United States, Europe, Japan, and Middle and Far Eastern countries by Indian gum manufacturers generally bears their own brand names. Most of the foreign, hydrocolloid companies, have now completely outsourced the production of their brand name of guar gum to some Indian or Pakistani guar gum manufacturers.

The first set of machines for use in the guar gum industry in India and Pakistan were imported from the United States and Europe. The best qualities of indigenous guar processing machines are now fabricated in India. It took very little time for the guar processing companies in India to master the original U.S. technology for making guar gum and to improve it.

The following flow sheet shows various steps in manufacturing guar gum.

Matured, dried, and cleaned guar seed (starting material)
↓
Course grinding; adjusted to produce guar split
↓
Unhusked guar splits and germ are produced
↓
Sieving and sifting to separate germ → Germ portion removed
↓
Unhusked split is heated in a rotary kiln at 120°C–150°C for about 45 seconds to loosen the husk
↓
Dehusking by controlled milling and sifting → Husk and any remaining germ are removed
↓
Purified and dehusked endosperm or the split is produced
↓
Split is mixed with an equal weight of water and flaked under heavy rollers to disrupt cellular structure
↓
Milling of flaked split into 100–300 mesh powder using hammer mill or ultrafine grinder
↓
Gum powder, thus produced, is sieved according to particle size
↓
Testing, quality control, and packaging of the powder for marketing

7.8   NEWER DEVELOPMENTS AND EXPORT OF PROCESSED GUAR PRODUCTS FROM INDIA[11]

The practice of exporting guar split and guar gum powder processed in India to Western countries started in the early 1960s. This practice has not only continued but has further increased. By processing of guar seed into the gum, the industry also gets its main by-products, which are the protein-rich germ and the seed husk (called *churi* and *korma*, respectively, in Hindi). A mixture of these two by-products has found use as much-needed animal feed for cattle in India. Besides being a major producer of guar crop, India is also a major cattle-breeding country. By processing guar seed into gum and cattle feed, the total bulk of the exportable guar polysaccharide (gum) is reduced to less than one-third of the original weight of the guar seed used, and this makes a considerable saving in the cost of product transport to countries that import guar gum.

Food-grade guar gum powder produced under good manufacturing practice and hygienic conditions by Indian companies meet internationally acceptable standards (e.g., Indian Standard Institution [ISI], U.S. Food and Drug Administration [FDA], EU Food Act) so that these gum products are permitted for sale and use in food products throughout the world. Besides the gum powder, some guar split is also exported to areas, including the United States, Europe, and Japan, where it is further processed into the so-called specialty guar products, which are still not produced to a significant level in India.

7.9 GUAR CROP AGRONOMY AND ECONOMICS[1,2]

Agronomy of the guar crop, its botanical aspects, and related topics has been dealt with in detail in Whistler and Hymowitz's monogram.[2] Besides this, an exclusive compilation of articles, "Guar: Its Improvement and Management," edited by Paroda and Arora,[6] has been published by the Indian Council of Agriculture Research. Hence, this topic shall be covered here only briefly. Those interested in plant sciences related to guar are referred to these excellent monograms for more details. The present book is primarily concerned with manufacturing technology, physicochemical properties, and industrial applications of guar gum and other galactomannans.

In India, the Central Arid Zone Research Institute (CAZRI) in Jodhpur, Durgapura Agriculture Research Station in Jaipur, Rajasthan Agriculture University in Bikaner (Rajasthan), and Hissar Agriculture University (Haryana) has been working on agronomy of guar crop. The Department of Chemistry at Jai Narain Vyas University and L. M. College of Science and Technology, both in Jodhpur, are the industrial research and development centers on guar products in India, where research on developments related to guar have been carried out. In the United States, T. Hymowitz[1] at the Agriculture Institute at Purdue University worked on a U.S. Department of Agriculture project to develop a high-yielding variety of guar seed, and R. L. Whistler (Whistler Center for Carbohydrate Research at Purdue University) collaborated with him in compiling information related to the chemistry and industrial utilization of guar gum. Hymowitz also visited CAZRI to collaborate on a project related to guar.

7.10 GUAR PLANT[2,6]

The guar plant is a pod-bearing, nitrogen-fixing annual legume. The plant grows to a height of 4 to 6 feet and bears a bunch of pods close to its main stem. Green pods are picked up and used as a raw vegetable or fried after drying. On ripening, the dried pods, now having turned yellow in color, are used for extraction of guar seeds.

The annual monsoon rain pattern (June to September) in the arid and semiarid northwestern region of the Indian subcontinent has been ideal for growing guar. Attempts to grow guar in the arid regions of the United States (Texas, Arizona, and Oklahoma) and certain other parts of the world were not very successful. Currently, India and Pakistan share 70% and 30%, respectively, of the total guar gum exported. Being a drought-resistant, hardy, and nitrogen-fixing legume plant crop, with low water requirement and no special requirements of soil, it is mainly grown in Northwestern India as one of the rain-fed crops.

In the past, when the crop was completely dependent on monsoon rains, there used to be large fluctuations in annual guar seed production and, consequently, the market price of guar gum. However, due to starting of guar crop production in certain irrigated areas in the states of Rajasthan, Gujarat, Haryana, and Punjab during the past 10 to 15 years, the supply and price structure of guar-based products has become more stable and dependable. An irrigated crop is reported to yield nearly twice (320 Kg/hector) the amount of seed as was earlier produced by a completely rain-fed crop per acre of land. Foreign exchange earnings by guar products exported by India is more than $40 million (Rs. 200 crore). Guar is a hardy crop, subject to very few diseases and pests. Matured and dried guar seed can be stored well for over one year and most insects do not infest stored guar seed.

7.11   MARKETING AND EXPORT[11]

In India, guar-producing farmers bring their guar seed produce to the grain markets, which are locally called *krashi mandi*, where the farm products are purchased by the traders and stored. Guar seed is processed in two stages and generally in separate industrial units. Guar split-making units are widely scattered in most of the guar-growing states of Rajasthan, Gujarat, and Haryana. Split-making units purchase guar seed from the local *krishi mandis*. These units can have an annual processing capacity of 1500 to 15,000 tons of seed into the split. The guar split produced amounts to slightly less than one-third of the total weight of guar seed processed, the rest being animal feed.

Consumption of guar gum in India is much smaller compared to its export. This is due to the fact that not many countries in the world produce guar gum, whereas it is being consumed all over the world as a food gum. Gum thickener for textile printing paste and oil-well drilling are two other important uses for which guar gum is exported, particularly to the developing countries dealing in textile printing and oil-producing countries.

Much of the marketing of guar gum and formulations based on it depends on the multinational companies dealing in a wide range of hydrocolloids, chemicals, and gums. Hence, it may not be very surprising if some guar gum exported from India is reexported to other world countries and even into India under different trade names, such as those of a textile thickener or as a petroleum drilling-mud additive.

7.12   BIOTECHNOLOGICAL ASPECTS

Biotechnology can be a useful tool in agronomy of guar and other crops. Presently the world's galactomannan gum market is particularly deficit in binary gel-producing galactomannans (e.g., LBG and tara gum), which are in great demand for the food industry. This gives rise to some interesting questions about the possibility of using biotechnology in guar agriculture and processing technology, which need to be answered. For example,

> Can gene technology help in producing galactomannan from an annual crop (i.e., guar) that will have a lower percentage of galactose, so that it can form binary gels?
> Can galactose grafts in guar polysaccharide molecules be partially removed, enzymatically, to bring down the M:G ratio to ~4:1?

These are the questions that cannot be answered satisfactorily, at least not now. Attempts to selectively reduce galactose grafts in guar gum using purified α-galactosidase enzymes too have not been very successful.

The Biotechnology Division of the Monsanto Co. (United States) has claimed to produce many genetically modified (GM) seeds of certain crops, for example, cotton, rice, maize, and wheat. But, so far no GM guar seed has been developed or introduced for commercial agriculture. Of course, improved non-GM guar seeds have been developed in many agriculture institutions and some universities in India. The farmers in India are now using these high yielding, non-GM guar seeds.

Already many of the foreign buyers of guar gum, who frequently visit India, are inquiring about any development related to GM seed of guar. Since the acceptance of GM grains for human food is not currently permitted in many countries, any use of GM guar seed (if produced in the near future) shall remain doubtful for agriculture. Indian manufacturers and exporters of guar gum are also asked to certify that their product is free from any pesticide residues and that no GM seed is being used from the guar seed procured to make a particular gum powder.

Recently a claim was made by biotechnology scientist Dr. Kanwarpal S. Dhugga[13] at Pioneer HI-Bred International Inc. (a subsidiary of DuPont Company USA) that the genes responsible for synthesis of galactomannans in carob and guar plants have been identified and these could be incorporated into the soybean plant. According to Dhugga, this could result in the soybean plant producing galactomannans, similar to those present in guar or locust bean seed and in the soybean seed. It has been argued that world production of soybean is not subjected to as much variations as those of the current guar seed produced on the Indian subcontinent. Even if galactomannan polysaccharide could be produced by a genetically modified soy plant, its industrial acceptability, particularly in food, remain doubtful, at least for some time to come.

With the technology of seed galactomannan extraction, there are problems likely to be encountered in separation of germ protein, endosperm polysaccharide, and seed oil when these are present together in a seed. An example of this is tamarind kernel powder (TKP), which contains all three constituents in sizable quantities, and the problem of effective separation of seed protein from the tamarind polysaccharide remains unsolved. This has also very much reduced the economical value of TKP. Thus, purified (protein freed) tamarind seed polysaccharide (Jellose), which is a good gelling agent similar to pectin, can attract those in the food industry and a better price if it could easily be separated from protein.

Another important fact overlooked while undertaking Dr. Dhugga's investigation has been that for the last 10 years or so, the agriculture of guar has also been carried out in many irrigated areas in India. This has to a large extent removed the uncertainty about variation in the annual crop output of guar seed, which in turn has considerably stabilized its export market, price structure, and supply position.

Two Japanese companies—Taio and Dinipon—have used biotechnology in selective and controlled enzyme depolymerization of guar gum to produce an ultra-low viscosity and a general-purpose dietary fiber from guar gum. In this process the mannan backbone of a guar gum molecule was selectively cleaved enzymatically using bacterial mannase. The M:G ratio in the product remained unchanged at ~2:1, which indicates that no galactose grafts were cleaved. Thus, there is much scope in producing more such enzymes to selectively modify guar gum, and for adoption of such enzyme technology by the guar gum industry in India. Biocon Co. (an biotechnology company in Banglore, India) has been working on the development of selective guar galactomannan modifying enzymes.

7.13 NUTRITIONAL ASPECTS OF GUAR SEED PROTEIN[2,14]

A research team, which is lead by this author, at Jai Narain Vyas University and another group of researchers at the Pharmacy Division of the L. M. College of Science and Technology are working on the modification of guar gum and utilization of guar germ protein for human food.

R. L. Whistler had earlier advocated developing technology for making guar seed protein to be suitable for human consumption. According to Whistler:[2] "In addition to enlarging of its [guar gum] present industrial uses, guar seed also holds great promise for supplying much-needed protein for human diet, in the developing countries."

Currently the guar germ (protein) and the husk obtained by milling of guar seed are mixed together and used as a protein-rich animal feed. During the early 1950s (before development of the guar gum industry), the coarsely ground, whole guar seed was used as feed for cattle in India. The galactomannan portion (endosperm) of guar seed is incompletely digested, even by ruminant animals, and it is partly excreted in cow dung.

After establishment of the guar gum industry in India, many cattle breeders, who lacked knowledge about animal nutrition, thought that the removal of the polysaccharide portion from guar seed shall reduce the nutritive component of guar-seed-based animal feed. This view was basically incorrect because the main carbohydrate component for ruminant animals comes from cellulose in their feed. It took some time and effort, through the print and electronic media, by the Indian government and animal nutritional experts to convince cattle breeders in India that the germ (protein) rather than the endosperm polysaccharide is the main animal nutritional component of guar seed. Amino acid composition of any protein being used as a component of human food or animal feed is an important criterion that determines the nutritional quality of the protein. Generally animal proteins, including milk, have a good balance of all the ten essential amino acids, which are not synthesized in the human body. Humans are wholly dependent on food as their source for these essential amino acids.

Plant proteins are generally deficient in their essential amino acids content. Compared to many other edible legumes, guar protein has a fairly good balance of nutritionally essential amino acids, including lysine and tryptophane. It has been reported that guar protein compares well in its essential amino acids content to soybean protein.

Certain antinutrients (e.g., trypsin inhibitors and lectins), which are generally associated with most legume proteins, are also present in guar seed protein.[15,16] These two constituents, which are proteins by nature, are completely denatured and rendered harmless on normal cooking of food. Hence, these should not produce any problem if guar protein is incorporated in the human diet. Another and more serious limitation of guar protein is its not very pleasant beanie odor. Simple dry heating treatment (roasting) removes the beanie odor from guar germ protein. About 4%–6% oil, which also contributes toward the unpleasant odor of guar protein, can be recovered from guar germ by solvent extraction. Though inedible, the oil thus recovered can be used for nonfood purposes. After roasting, guar germ develops a likeable taste, similar to certain other roasted legume seed powder (e.g., Bengal gram [roasted, *channa*]), which is eaten as snacks in India and other Eastern countries.

Technology for texturing soybean protein for human food was developed in Japan and China. This technology can also serve as a model for the development of textured guar protein for human food. Textured guar protein can be used, mixed with other vegetable proteins, to make protein-rich nuggets for human consumption similar to those that are already being made from soy protein. Thus, guar seed can contribute significantly to increase the availability of vegetable protein for human dietary needs in India and other developing countries.

When this aspect of guar protein was brought to the attention of the United Nations Organization (UNO) and World Health Organization (WHO), a request was made for a document to provide known facts about the potentials of the guar crop and use of its protein, resulting in publication of a booklet by Whistler and Hymovitz.[2] Unfortunately no attempts have so far been made in guar-producing countries for better utilization of the protein component.

7.14 CHEMICAL AND STRUCTURAL STUDIES OF GALACTOMANNANS[10,15,17,18]

With the realization that guar gum was soon to emerge as an industrial polysaccharide, extensive study of its chemical structure was undertaken by Whistler's group[18] of carbohydrate research. Prior to this, at the University of Edinburgh, Hirst and Jones, and Smith, in the year 1948, carried out chemical structural studies of LBG, which were based on exhaustive methylation of the polysaccharide, followed by hydrolysis of the fully methylated product. These methylation-hydrolysis methods of studying the chemical structure of polysaccharides were earlier used for cellulose and starch, and these formed a basis of similar studies of guar and all other galactomannans.

The chemical structure of guar gum was established by the classical work of Whistler's group, in the year 1950, at Purdue University, and it was based on the standard methods of polysaccharide-structure determination methods. These methods consists of

1. Determination of monosaccharide composition of the guar galactomannan.
2. Methylation-hydrolysis study, for determining mode of linkage of constituent sugars.
3. Partial enzyme hydrolysis to determine the mode of stereochemical linkage of various sugar components.
4. Molecular weight range determination by size exclusion chromatography and several other methods applicable to high molecular weight soluble polymers.
5. Periodate oxidation to determine the site and extent of branching.

Configuration at the intersugar monomer bond in galactomannans has been studied using enzymes.

Several improvements in the methodology for structural studies of polysaccharides (e.g., the chromatographic methods for separation and estimation of component sugars) and their derivatives, nuclear magnetic resonance (NMR) spectroscopic methods, and other physicochemical structural study methods for carbohydrate research have been developed since the 1960s. These and earlier structural chemical studies on carob galactomannans (LBG) served as a model for the structural studies of most other galactomannans.

Methods of polysaccharide structural study have been simplified and improved with the introduction of chromatographic methods (gas liquid chromatography [GLC] and high-performance liquid chromatography [HPLC]) of sugar analysis, NMR, including ^{13}C, and mass spectral methods and improved methods of polymer molecular weight determination. It has been estimated that the molecular weight of undegraded galactomannan molecules in guar gum is in the range of one million (10^6) dalton.

These chemical structural studies have indicated that guar gum and most of the other legume galactomannans have some common structural features. Their molecules are built up of a linear backbone of $\beta(1\rightarrow4)$-linked mannopyranose sugar monomers. The linear, mannan polymer backbone is substituted by single galactose grafts attached to the mannose unit at the C-6 position by $\alpha(1\rightarrow6)$ linkage.

Figure 7.1 represents an idealized fragment of guar galactomannan molecule, where the M:G ratio is presumed to be 2:1 and the positions of linkages and configurations of the intersugar bonds is shown.

Thus, from all these studies, it has now been concluded that guar gum is a linear galactomannan, the molecule of which is composed of $\beta(1\rightarrow4)$-linked mannopyranose backbone, with numerous branch points (grafts) from the C-6 position of

----(1→4) - β- D- Man- (1→4) - β- D- Man- (1→4) ----

↑

(1→6) α-D- Gal

FIGURE 7.1   An idealized representation of a fragment of guar gum.

mannopyranose, linked by $\alpha(1\rightarrow6)$ bond to a single D-galactopyranose sugar. There are between 1.5 to 2 mannose (backbone) units for every galactose unit (graft). The average M:G ratio in guar gum molecule being 2:1 or more accurately 1.85:1.

Periodic acid (HIO_4) oxidation of the polysaccharide guaran molecules has also suggested it to be a highly branched polysaccharide. The formation of a strong, flexible, and stretchable guaran triacetate derivative, which becomes anisotropic to polarized light, has further suggested it to be a linear polymer with numerous short-side chains or grafts.

7.15 FINE STRUCTURE OF GUAR GALACTOMANNAN[10,18,19]

The model of the galactomannan molecule fragment (idealized structure) does not represent the true picture of guar gum. According to an earlier suggestion by Whistler and BeMiller,[10] the single galactose sugar grafts on the mannan molecular backbone were uniformly placed on the alternate mannose sugars of the backbone. A detailed study of distribution of these grafts, which is based on the controlled enzymatic hydrolysis and identification of the oligosaccharides produced, has now shown that the placement of single galactose grafts is random. It follows that small continuous blocks of two to six galactose sugars may be present at one site on the mannan chain, and blocks of bare backbone separate these. It has also been established that doublets of graft and not triplets are of a common occurrence.

There is a certain degree of freedom of rotation, along $\beta(1\rightarrow4)$-linked glycoside oxygen joining two adjacent mannose units of the galactomannan polymer backbone. When the freedom of such a rotation is reduced in certain polysaccharides (e.g., cellulose), its molecules tend to have a straight, ribbonlike conformation. To a certain extent, such restriction to rotation could also arise due to intrachain hydrogen bonding of two adjacent mannose units in a mannan polysaccharide. With a large number of the galactose grafts on a galactomannan polysaccharide, the possibility of hydrogen bonding of two adjacent mannose of the backbone is reduced. This produces a structure like a garland of leaves. In this structure, the leaves, joined at the ends, can rotate and therefore these can orient in different ways. Such freedom of rotation is reduced to some extent, in case of lower galactose substituted galactomannans (e.g., LBG), due to hydrogen bonding. In the case of a cellulose molecule, there is almost no freedom of rotation for glucopyranose units because it has no grafts at all.

Galactomannans from different legume seeds have different molecular weight and M:G ratio as well as the mode of distribution of galactose grafts along the main chain. In general, it was thought that the galactose grafts have a regular distribution. Thus, for guar gum the galactose grafts were thought to be present on alternate mannose (M:G = 2:1), whereas for LBG, every fourth mannose was supposed to carry a graft. This however has proven to be wrong and the galactose grafts are now supposed to have a random distribution or be present in blocks (called the *hairy region*), leaving portions of backbone unsubstituted or smooth regions (nonhairy). To a large extent the functional properties of the galactomannans are determined by the size of these substituted blocks and the unsubstituted regions separating them. Thus, in case of LBG the unsubstituted mannose blocks can be as large as 20 mannose units or more. These bare blocks show strong chain–chain interaction with molecules of its own kind as well as with those of other polysaccharides. Galactose grafts in guar gum are generally present as doublets and triplets, whereas the smooth blocks are generally less than six mannose units. Thus, chain–chain interactions for guar gum are much weaker compared to those in LBG.

Since a large number of grafts on the polymer backbone reduce the chance of chain–chain interaction, the polysaccharide molecular chains with a higher percentage of galactose have less chance of coming close and from an interchain hydrogen bond. Hence, these galactomannans with more galactose disperse readily to a molecular level in solution and are cold-water soluble.

Each molecule of guar gum has the size of a colloidal particle. Guar gum solutions exhibit very high viscosity. Viscosity of a polysaccharide also depends on its molecular weight and its concentration in a sol. Earlier guar gum was thought to have a molecular weight in the range of ~500,000, but later estimates have shown it to be in the range of ~1–2 × 10^6 dalton.

One important feature of the monosaccharide components (i.e., mannose and galactose in guar gum and other galactomannans) is that these sugar units each have a pair of *cis*-pairs of hydroxyl groups at C-2, C-3, and C-3, C-4 positions. Because of the presence of these *cis*-pairs of hydroxyl groups, galactomannans can form strong (three-center) hydrogen bonds with other polysaccharides, and get adsorbed on hydrated surfaces, for example, those in clay and cellulose. Hence, galactomannan polysaccharides are referred to as hydrogen-bonding type reagents.

7.16   SOME PHYSICOCHEMICAL PROPERTIES OF GUAR GUM[10]

The methods for studying physicochemical properties and structural study for polysaccharides essentially consist in preparation of an analytically pure polysaccharide sample by repeated alcohol precipitation from a clarified aqueous solution. This is followed by determination of its average molecular weight and percentage distribution of molecular weight in a polysaccharide sample. This is done by earlier known as well as some later developed methods.

Good quality food-grade guar gum is a white powder available in different mesh sizes (150–300 standard U.S. mesh). Gum powder is soluble, or rather dispersible, in cold water and produces a highly viscous, translucent paste, which can have an apparent viscosity in the range of 2000 to 8000 cps when fully hydrated as measured by a Brookfield viscometer (Spindle No. 3, at 20 rpm) at 25°C.

Commercial guar gum powder and its paste turn yellowish on addition of aqueous sodium hydroxide at and above pH 8. This indicates the presence of some (traces) of a pigment in guar gum, which acts like an acid–base indicator. In lab samples of guar gum, this pigment can be removed by repeated washing of gum powder with an aqueous–alcoholic caustic, followed by neutralizing and washing with alcohol. Such treatment yields powder, freed from pigment, which does not turn yellow at alkaline pH (>9.0).

The optical rotation ($[\alpha]_D$) of guaran (pure guar gum), measured in 1.0N NaOH, has been found to be 53° at 25°C.

TABLE 7.2
Typical Composition of Commercial Guar Gum

Component	Percentage (%)
Galactomannan (by difference)	75–85
Moisture (loss on drying at 105°C)	8–12
Proteins (N × 6.25)	3–6
Insoluble fiber (cellulose)	2–3
Ash (minerals)	0.5–1.0

Guaran is made of nonionic, polydisperse, rod-shaped molecules. It can be fractionated into portions having a variable M:G ratio (in a range 1.5:1 to 3:1) and having different water solubilities. Solubility of guar gum fractions having lower percentages of galactose is less. The size of the rodlike guar molecule is much larger in length compared to the LBG molecule. On average a guar gum molecule is built up of about 10,000 residues of D-mannopyranose and D-galactopyranose taken together. Higher galactose substitution on the mannose backbone in comparison to LBG increases the stiffness of the molecule and decreases its flexibility. It also reduces the overall extensibility and radius of gyration of its isolated molecular chains.

It has been reported that intra- and interchain hydrogen bonding between linear galactomannan polysaccharide molecules can result in time-durable junction zones. This requires a minimum of continuously unsubstituted, six or more mannose groups (nonhairy portion) on the molecular backbone. In case of guar gum, more frequent galactose grafts on the mannan backbone reduces strong chain–chain hydrogen bonding interactions and improves its aqueous solubility. Since large, unsubstituted portions of backbone are not of frequent occurrence on guaran molecular backbone, it has good water solubility among various plant galactomannans.

A typical composition of a commercial sample of guar gum powder, as reported in an analysis report of a manufacturer is shown in Table 7.2.

7.17 MANUFACTURE OF GUAR GUM FROM SEED[10,11]

Good quality guar seeds are large, smooth and rounded, and weigh ~780 g/l. The quality of guar seed produced depends on the quality of the seed used for its crop in addition to the regional conditions (e.g., soil), and trend of rainfall or irrigation. More rainfall after the seed has matured on the plant can darken the seed and reduce the gum yield and lower its quality. With an irrigated plant, unless the watering is reduced or stopped after a certain time, the growth of the plant continues but it produces more foliage rather than good quality seeds. This sometimes also happens to the rain-fed crop due to the long and well-distributed but prolonged monsoon season.

The bulk of guar seed arriving at any agricultural produce market (*krishi mandi*) represents the average quality of the seeds from different growing regions brought

TABLE 7.3

Composition of Guar Seed and Its Components

Seed Part (%)	Composition (%)					
	Protein	Lipid	Ash	Moisture	Fiber	Galactomannan
Hull (14–18)	3–5	0.3	3–4	8–10	33–35	1–2
Gum (33–44)	2–3	1–2	10	6–10	2–3	70–80
Germ (40–45)	50–55	5–6	4–5	8–10	15–18	5–6

to a particular marketing place. Average composition of guar seed is shown in Table 7.3.

According to the current general practice, manufacturing of guar endosperm split (*dal*) is carried out in exclusive, split-making units. These units produce and sell guar split to the stockpiles in those places where units that make guar gum powder are located. The stockpiles in turn supplies the split to the guar powder-making units as well as to the exporters of guar split. The by-product, germ portion mixed with the husk, which forms nearly 70% of the guar seed weight, is sold as animal feed. There are very few composite units making both guar split and powder.

7.18 PRODUCTION OF GUAR SPLIT[2,10]

On arriving in a guar split-making unit, guar seed is screened to remove extraneous material using standard seed-cleaning vibrators, electromagnets, and sifters. Cleaned seeds are split into endosperm halves by a parallel plate grinder, one plate of which is stationary and the other rotating at 1500–1800 rpm. The clearance between the two plates is so adjusted that less than 5% of seeds escape splitting and a minimum amount of splits are further broken into pieces. The germ portion of the seed, which gets detached from the endosperm half, being soft, gets powdered. The husk or the seed coat remains on the endosperm at this stage. The germ particles are separated from the split at this stage by differential sieving and sifting.

The seed coat (husk) carrying endosperm splits are then heated in an oil-fired, inclined, rotatory kiln at a temperature of 105°C–115°C for ~90 seconds to loosen the seed coat, which now becomes brittle. The endosperm remains resilient and rubbery at this stage. Dehusking is now easily carried out in a dehusking machine, which consists of a two-tiered chamber having rotating saw-toothed blades. During dehusking care has to be taken to control the temperature and time of heating (in the kiln) of the husk attached to the split, because excessive heating can cause browning of the split and produce low viscosity guar gum. The seed coat, rendered brittle by heat, easily gets powdered and passes through, while the split is retained on a 10-mesh size sieve. Some broken splits are also recovered, but these yield gum powder of lower grade. Dehusking of better than 5% of the (unhusked) splits is generally achieved. Dehusked split is graded according to the percentage of undehusked split still present with it, and heat browned split, green or blue split (from unripened seed) present, in a sample. A good quality of light yellow split contains on a dry basis >93% galactomannan.

7.19 PRODUCTION OF GUAR GUM POWDER FROM GUAR SPLITS[2,10,11]

Grinding of dehusked split to gum powder is carried using a closed, ultrafine attrition mill or hammer mill pulverizers to make guar gum powder. For high viscosity and fast hydrating product, the split is mixed with an equal amount of good quality soft or preferably deionized water, in a rotating, double-cone blender (50–200 Kg capacity). Certain chemicals, for example, buffering and whitening agents (acetic acid and hydrogen peroxide) and preservatives (sorbates and benzoates) are optionally added to water at this stage. The splits swell upon absorbing up to their own weight of water and they can then be flaked under high pressure using hydraulically operated, heavy, rubber- or ebonite-lined steel rollers. This treatment disrupts the cellular structure in the seed endosperm. According to one patent, when the clearance between the flaking rollers is about 30 microns, more effective disruption of cellular structure takes place, resulting in high viscosity and fast hydrating gum powder.

Flaked splits continuously fall into a large hopper from where it is sucked into the impact pulverizer and gets powdered. To reduce any depolymerization due to the heat generated during this process, completely closed and multichamber pulverizers, which are locally known as ultrafine grinders are employed. During grinding, the air contact is thus minimized. Some moisture is lost by evaporation during the grinding and this reduces the frictional heat generated in grinding and keeps the temperature low. This in turn reduces excessive heat depolymerization of the polysaccharide during grinding. The powder coming out of the ultrafine grinder still contains ~30% moisture. Finally this powder is flash dried in hot-air driers and sieved into different particle sizes. Energy consumption of such processes is very high.

In case of a hammer mill pulverizer, the powder coming out contains only 8%–10% moisture and it does not flash dry. In general the powder produced by an ultrafine machine has a higher viscosity compared to that produced by a hammer mill.

Composition of typical food-grade guar gum was shown in Table 7.1. It will be seen that commercial guar gum powder is highly purified galactomannan and contains >90% galactomannan on a moisture-free basis.

7.20 MODIFIED GUAR GUM PRODUCTS[10]

It was earlier mentioned that worldwide increased demand for guar gum has resulted due to the manufacturing of its specialty products, that is, the derivatives and modified guar gum products. For modified guar gum products, the basic galactomannan structure does not change, but the molecular weight, viscosity, solubility, rate of hydration, and so forth can be altered by hydrolytic or oxidative depolymerization. Such products can be used in food unless blended with nonfood constituents. A large number of acid, hydrogen peroxide, hypochlorite, and enzyme modified guar products have been commercialized for specialized applications. Controlled depolymerization of guar gum has also been carried out by thermocatalytic method.

In enzyme depolymerization, the linear mannose backbone of a galactomannan is selectively cleaved by a purified, bacterial mannase enzyme. If desired the galactose grafts can also be cleaved, selectively, using purified galactosidase enzymes, though this has not been very successful on commercial scale, to produce LBG-like products. Cleavage of galactose side chains does not produce a large change in the solution viscosity, but backbone cleavage, by mannase enzymes, cause drastic lowering of viscosity.

Germinated seeds of a guar also produce these hydrolytic enzymes. These mixed mannase and galactase enzymes are known to breakdown guaran nonselectively. For selective depolymerization only the purified bacterial enzymes, which are produced commercially, have been used for modification of guar gum on an industrial scale.

Certain galactomannan hydrolyzing enzymes were successfully produced for the first time by the Indian biotechnology giant Biocon of Banglore city in Karnataka state. Mannase enzyme produced by this company have been used by this author, and these were found to be as effective as those being used by Japanese companies to produce ultralow viscosity guar gum.

Two Japanese companies, namely, Dainippon and Taio Kagaku, have produced enzyme depolymerized, ultralow viscosity guar gum for use as food fiber. These low viscosity guar gum products have been approved by the U.S. FDA as safe food additives and are being marketed under the trade names Fiberon, Sunfiber, and Benefiber. The Indian company Lucid Colloids, under a technology transfer agreement with Taio Kagaku, is now using Japanese enzyme depolymerization technology to produce ultralow viscosity Sunfiber in India. Since the insoluble portion is completely eliminated from ultralow viscosity depolymerized guar gum, it produces a completely colorless transparent, water-white aqueous solution. Such low viscosity and completely odorless and tasteless guar gum powder can be added to health beverages and foods without any change in the texture of the original product and provide an adequate amount of soluble dietary fiber.

Normal guar gum powder has a faint, beanie odor. More recently, there has been an increasing demand and inquiries from importers of guar gum from India about the availability of completely odorless guar gum with improved transparency. This is sometimes achieved by expelling volatile, odor-producing constituents by steaming or washing the guar split with dilute alkali, followed by neutralizing and alcohol and water washing. Prior washing of the split to remove proteins and other insolubles improves transparency of the gum thus produced.

Delayed viscosity developing and dispersible or nonlumping guar gum varieties have been formulated by controlled borax cross-linking at alkaline pH, and blending it with solid fumeric acid. After dispersing of such a gum formulation into water without lumping, the fumeric acid present causes neutralizing of alkaline pH in the aqueous solution. This results in delayed development of viscosity.

Delayed viscosity development of guar gum has also been achieved by cross-linking with glyoxal. A bis-acetal is produced by this dialdehyde, resulting in cross-linking, at pH ~3–4, which also produces dispersible gum powder. In this case the cross-linking is reversed on raising the pH >8 when the acetal bonds are broken, causing guar gum to dissolve. For more details, see Chapter 4, where glyoxal cross-linking is shown in Figure 4.2.

7.21 GUAR GUM DERIVATIVES[10]

In contrast to modified guar gum products, derivatization of any polysaccharide involves chemical reactions, which are mainly etherification and less frequently esterification of the hydroxyl group in a galactomannan molecule. This results in altering its basic galactomannan structure, particularly at the site of the hydroxyl group. Any derivatization process involves the cost of the reagents used and the process cost. Derivatization of only cheaper polysaccharide materials (e.g., starch, cellulose, and guar gum) has been economically feasible and has been practiced. Many patents on derivatization of galactomannans other than guar have been reported in the literature, but these are not as yet practiced commercially. This topic of derivatization of polysaccharides is covered in Chapter 5.

On an average, there are three free hydroxyl groups on each sugar residue in a polysaccharide molecule, which can be substituted, giving a maximum degree of substitution (DS) as three. In actual practice the DS is seldom kept more than one. Dramatic change of properties of a polysaccharide can even be observed at as low DS as 0.1. Etherification using monochloroacetic acid (MCA) or its sodium salts (SMCA), methyl chloride, and ethylene or propylene oxides has been the most common derivatization reactions for polysaccharides. Anionic, cationic, and nonionic derivatives of guar polysaccharide that are most commonly produced and used are shown in Table 7.4.

Besides these ether derivatives, many more guar ethers, for example, alkyl, allyl, and those having thiol and amino group (side chain) substituted products have been reported. Many of these are not yet produced commercially. Guar esters of some organic and (acetic) inorganic acids were also prepared, but only phosphate ester is reported to be prepared commercially.

Etherification with MCA or CH_3Cl proceeds by standard Williamson's etherification protocol. Polysaccharide etherification is a heterophase reaction yet a reaction efficiency of better than 70% is easily achieved.

Phase transfer catalysts were successfully used in these heterophase reactions for the first time by the author, resulting in improved reaction efficiency and the product yield. In the reaction of sodium monochloroacetate, one mole of alkali is used in the reaction to neutralize the acid produced in the reaction. The reaction

TABLE 7.4
Currently Manufactured and Marketed Guar Gum Derivatives

Derivative	Substituent Group	Charge	DS or MS
Carboxymethyl (CM)	$CH_2COO^- Na^+$	Anionic	0.1–0.5
Hydroxypropyl (HP)	$-CH_2.CH(OH).CH_3$	Nonionic, hydrophobic	0.3–2.0
2-Hydroxypropyl-3-N-trimethyl ammonium chloride (Quaternary)	$-CH_2.CH(OH). CH_2-^+N(CH_3)_3Cl^-$	Cationic	0.05–0.3
CM,HP Double derivative	$-CH_2COO^- Na^+$ $-CH_2.CH(OH).CH_3$	Anionic, hydrophobic	0.05–0.2

presumably proceeds via a polysaccharide alkoxide formation. In case of alkene oxides etherification, only a catalytic amount of alkali is required. These reactions are shown next.

$$PS\text{-}OH + NaOH \rightarrow PS\text{-}ONa + H_2O, \text{ where } PS = polysaccharide$$
$$PS\text{-}ONa + ClCH_2COONa \rightarrow PS\text{-}O\text{-}CH_2COONa + NaCl,$$

and

$$PS\text{-}ONa + CH_2\text{-}CH_2\text{-}R \longrightarrow PS\text{-}O\text{-}CH_2\text{-}CH\text{-}R$$
$$\diagdown \diagup$$
$$O$$

Guar gum etherification reactions were developed on the same line as those developed earlier for cellulose and starch. These reactions can be carried out on a lab scale in an aqueous slurry containing a nonsolvent, for example, 2-propanol. On a commercial scale, it is economical to carry out the reaction on hydrated guar splits or powder containing its own weight of water. Since lumping can take place when water and the aqueous solutions of the derivatization reagents are not sprinkled properly, it is always better to use guar splits. Reacted splits can later be washed to eliminate the reaction by-products (NaCl, excess NaOH, and hydrolyzed MCA). Flaking and grinding of the derivatized splits follows the reaction.

In a hydroxyalkylation reaction a new hydroxyl group is present on the newly formed side chain (-$CH_2.CHOH.CH_2R$), which can become the site of further reaction. This can result in lengthening of the side chain. In such cases the substitution is expressed in terms of molecular substitution or MS. A typical side chain having MS = n + 2, is shown next.

$$PS\text{-}O\text{-}CH_2.CHO.CH_2R$$
$$|$$
$$(CH_2.CHO.CH_2R)_n$$
$$|$$
$$CH_2.CHOH.CH_2R,$$

where R = H, -$^+N(CH_3)_3Cl$, -SO_2ONa.

In most of these etherification reactions, the reactivity of the various hydroxyl groups in a monosaccharide unit follow the order C-2 > C-6 > C-3. Overall, the C-2OH (secondary) and C-6OH (primary hydroxyl) are the preferred positions for substitution.

Derivatization by etherification of guar gum increases room temperature solubility and overall transparency of the aqueous solution of a derivative. The rate of hydration of the resulting derivatized gum product is also improved. Hydroxypropyl or HP derivative of guar gum has much improved compatibility with water-soluble organic solvents (e.g., lower alcohols) and has much better solid suspending ability. Thus, HP guar gum is soluble in 50% aqueous methanol. With increasing DS, the products become less prone to microbial attack and thermal stability is improved. Because of the improved functional properties, guar HP derivatives have found extensive applications in oil fields, for example, deep drilling of oil wells and fracturing of an existing oil well for increasing oil and gas recovery.

By periodate oxidation of guar gum some *vic*-hydroxyl group pairs are selectively cleaved to produce aldehyde groups. Such periodate oxidized or the aldehydic guar gum derivative has also been prepared for specific applications, for example, for use in paper sizing.

7.22 TESTING AND QUALITY CONTROL OF GUAR GUM PRODUCTS

Particle size of any commercially produced guar gum powder is generally graded according to sieve analysis (U.S. standard sieve). Generally, 100 to 300 mesh size powder is marketed. In each case, the percentage of powder passing through a specified sieve size is reported.

Additionally the apparent viscosity (in centipoise [cps] units) as measured by a rotary viscometer (e.g., various models of Brookfield viscometers) and the pH of 1.0% w/w, aqueous solution are reported. Residual protein content (based on nitrogen determination), ether extractable (lipids), ash (minerals), and heavy metals (Pb, Hg, Cd, etc.) present in a gum sample are also reported.

Microbial analysis (in the case of food grade products) is carried. This is reported as the total plate count and any other pathogens, which might be present in a gum sample. In food-grade guar gum manufacturing and exporting units in India, a well equipped microbial testing lab (sterilized area) and properly trained staff are a necessity. Microbial testing reflects on the stability of the powder and its aqueous pastes.

Depending upon the processing method, gum powders of viscosity ranging from 4000 to 8000 cps and up to >90% passing through a specified standard sieve (150–300 U.S. mesh size) is generally reported. These quality control labs of the guar gum manufacturers in India are frequently inspected by government agencies.

Typical and desirable composition of food-grade guar gum is shown in Table 7.5.

The author has developed methods for determination of DS and MS in guar gum derivatives and these are being used in some industrial guar gum product testing labs

TABLE 7.5

Typical Composition of Guar Gum as Reported by the Manufacturer

Specification of Guar Gum	Composition (%)
Moisture	8–12
Protein (N 2.6)	3–5
Lipid	0.3–0.6
Acid insoluble residue	1.5–3.0
Ash	1.0–1.5
Galactomannan (by difference)	75–85

in Jodhpur when needed. Determination of acid insoluble residues (AIR) and filterability tests (through an appropriate textile printing screen) is generally reported for products meant for the textile printing industry.

For export purposes, the powder is generally packed in 25–50 Kg lots into polyethylene-lined multiple-ply paper bags. When properly stored, the product gum can have a storage life of over one year without appreciable change in viscosity.

Specifications of food-grade guar gum are given by the ISI, Food Chemical Codex, WFO, WHO, and FDA in addition to various pharmacopoeias. Guar gum meant for food and pharmaceutical applications should be free from preservatives or may contain only allowed food preservatives.

7.23 RHEOLOGY OF GUAR GUM PASTES[10,16]

Viscosity is the primary denominator to judge any guar-gum-based product. It reflects on refinements of the method used for manufacturing guar-gum-based products. Like most of the high molecular weight, linear polysaccharides, guar gum solutions at concentrations >0.2% start showing non-Newtonian or shearing thinning flow behavior changing to thixotropic behavior at concentrations >1.0%.

Guar gum, because of its very high viscosity, is an economical and preferred thickener and stabilizer for food and nonfood industries. It hydrates fairly rapidly in cold water to give highly viscous, pseudoplastic solutions of high viscosity at low shear. Viscosities of guar gum solutions are higher when compared to most other plant hydrocolloids at the same concentration. Thus, the viscosity of a guar gum solution is much higher compared to that of LBG or any other galactomannan solution at the same concentration.

Being cold-water soluble, guar gum is a better emulsifier when compared to LBG. Higher solubility of guar gum is due to its having more galactose grafts on mannan backbone of its molecule than LBG. Unlike LBG, guar gum paste does not form weak gels upon cooling and its sol shows good stability to repeated freeze–thaw cycles. Guar gum sols, which show high viscosity at low shear, are strongly shear thinning at high shear rate, for example, on stirring at high revolutions per minute. Being a nonionic polysaccharide, the viscosities of guar sols are not much affected by the presence of neutral salts (ionic strength) or pH in the range 6 to 9, but it will degrade at very low pH and higher temperature (pH ~3 at >50°C). It shows viscosity synergy with xanthan gum. With casein, it becomes slightly thixotropic forming a biphasic system containing casein micelles.

Under laboratory conditions and keeping depolymerization of galactomannan during commercial extraction to a minimum, a guar gum powder can produce a high, limiting apparent-viscosity of 7500–8500 cps, as measured by a Brookfield viscometer, when it is fully hydrated at 1% w/w concentration in water. Some manufacturers of guar gum have claimed to produce guar gum of viscosity much higher than this upper limit of 8500 cps. Such claims are either exaggerated or the viscosity enhancement is done by synergic blending of guar gum with other polymers (e.g., xanthan) or by very low cross-linking.

Upon applying a shear (stirring) to a guar gum solution, the entangled random-coil molecules of guar galactomannan tend to get disentangled and the molecules get oriented in the direction of flow. This results in lowering of intramolecular friction and hence the solution viscosity is decreased. Hence, it is necessary to mention the shear rate at which the viscosity of a guar gum sample has been measured at a given temperature and concentration.

According to the current practice, most of the manufacturers of guar gum report the viscosity value of their products, as measured by a Brookfield viscometer (RVT or other models) at 25°C, using Spindle Number 3 and at 20 rpm. Like most polymers the viscosity of guar gum solutions decrease with a rise in temperature. The apparent viscosity at a given shear rate is reported in centipoises per second or the centipoise units. The viscosity values reported in this way help in determining the applications of a particular gum product. Graphically it has been possible to extrapolate viscosity on a gum solution to zero-shear value. In frozen foods (e.g., ice creams) guar gum retards ice crystal growth, nonspecifically, by slowing down mass transfer across the solid–liquid interface.

7.24 FACTORS INFLUENCING VISCOSITY OF GUAR GUM SOLUTIONS[16]

It was mentioned that the viscosity is the primary determinant to judge any guar gum product, which also reflects on refinements of the method for manufacturing guar-based products. High molecular weight, linear polysaccharides (e.g., guar gum solutions) start showing non-Newtonian or shearing thinning flow behavior when the concentration is >0.2% w/w. Guar gum solution viscosity changes to thixotropic behavior when the concentration is higher than 1.0% w/w.

Under laboratory conditions, and by keeping depolymerization of guar galactomannan during its commercial extraction to a minimum, a guar gum powder can produce a high, limiting apparent viscosity of 7500–8500 cps when it is fully hydrated at 1% w/w concentration in water.

Some manufacturers have claimed to produce guar gum of a viscosity much higher than this upper limit (8500 cps). Such claims are either exaggerated or the viscosity enhancement is done by synergic blending of guar gum with other polymers (e.g., xanthan) or by very low cross-linking.

On applying a shear to a guar gum solution, the entangled random-coil molecules of guar galactomannan tend to get disentangled and the molecules are oriented in the direction of flow. This results in the lowering of intramolecular friction and hence the solution viscosity is decreased. Hence, it is necessary to mention the shear rate at which the viscosity of a guar gum sample has been measured at a given temperature and concentration.

7.25 SUPERENTANGLEMENT OF GUAR GALACTOMANNAN MOLECULES[20]

Thickening behaviors of galactomannans show many similarities to other (β1→4)-linked linear polysaccharides, for example, carboxymethylcellulose (CMC), glucomannan, amyloids (tamarind gum), and xylanes. Thus, the double logarithmic plot of zero-shear viscosity against concentration multiplied by the intrinsic viscosity for these polysaccharides gives a straight line plot with a break, which corresponds to the transition from dilute solution behavior (i.e., when there is a low degree of coil–coil overlap) to a concentrated solution behavior. Transition from dilute to concentrated solution occurs when there is complete overlap and interpenetration of the polymer coils.

In case of guar gum there is an early setting of this concentrated solution behavior, which has substantially larger concentration dependence. Earlier setup of this concentrated solution behavior indicates that the coil–coil entanglement in case of guar gum is much stronger and more time-dependent when compared to most other linear polysaccharides.

This phenomenon, which is also observed for other galactomannans, has been referred to as *superentanglement* (entanglement between coiled linear polymer strands of molecules). Superentanglement of coils accounts for much higher efficiency of guar gum as a thickener when compared to many other linear polysaccharides. In the opinion of this author, superentanglement arises due to strong hydrogen-bonding interactions of galactomannan molecules and this has been attributed to the presence of *cis*-pairs of hydroxyl groups in sugars groups on the guar polymer backbone. The *cis*-pair of a hydroxyl pair can form three-center hydrogen bonds, which are more durable on a time scale.

7.26 SOLUBILITY AND HYDRATION RATE OF GUAR GUM[10]

Though guar gum is referred to as a cold-water soluble gum, it gives an opalescent viscous solution in a 1% aqueous dispersion and hydrates fully in about 2 hours, which is indicated by the maximum viscosity development. Hydration of the gum is fast at pH 6.0–8.5, and the hydration rate decreases on lowering or increasing pH from this optimum range. High molecular weight guar galactomannan polysaccharide molecules form a colloidal dispersion in water, but some cellulosic and protein impurities (~3%) present in the commercial guar gum samples remain suspended and these insolubles produce opalescence in the sol.

The average M:G ratio in guar gum is approximately 2:1. In reality the gum contains fractions of variable M:G ratios, of which a small fraction of insoluble galactomannan has an M:G ratio as low as 6:1. Fractions of guar gum having a very low percentage of galactose remain insoluble in water. However, such insoluble portions of the gum remains suspended, when the overall viscosity of guar gum solution is high. In gum solutions at lower concentrations of 0.1%–0.2% (i.e., when the viscosity is very low) these insoluble portions have a tendency to settle down. These can be filtered off or centrifuged to give a water-clear supernatant solution.

The rate of gum hydration, as indicated by the time taken to attain the maximum viscosity, depends on the particle size and the method of its manufacturing. Thus, it has been possible to manufacture products having different hydration rates that will produce the same, ultimate viscosity on complete hydration, but in different times.

Being nonionic in nature, guar solutions are compatible to other natural polysaccharides, proteins, and synthetic polymers, which may be ionic or neutral. Guar solutions show a strong synergy in viscosity (increase, which is higher than that predicted as the sum of its hydrocolloid components) with other linear polymers, particularly the anionic ones (e.g., CMC, xanthan, alginates, and synthetic poly-acrylates). Due to cross-linking, guar solution increases viscosity in a presence of borate ion at alkaline pH (>9) and can form a gel when borate concentration is higher. Guar solutions when stored for longer duration can reduce in viscosity due to bacterial decomposition.

The following empirical equation has been suggested to evaluate the approximate hydration rate of fine mesh (>250) guar powder as measured by the Brookfield RVT viscometer at 20 rpm at 25°C:

$$[\eta] = 452 + 656 \ln T,$$

where [η] is the apparent viscosity, and T is the hydration time, up to 120 minutes.[16]

An empirical relation has also been derived between viscosity and concentration for the course (<150 mesh) and fine mesh (>250 mesh) guar gum powders. According to the following equations, the viscosity can be computed by the relation

$$[\eta] = 3104 \times [C]^{3.11}$$

for the fine-mesh powder and

$$[\eta] = 2274 \times [C]^{3.27}$$

for course-mesh guar gum powder. In these equations, C is the concentration and [η] is the apparent viscosity at 25°C. These equations show that viscosity increases 8- to 10-fold when concentration is doubled.

An empirical relation between the molecular weight of guar gum and its intrinsic viscosity has also been worked out, which is as follows:

$$[\eta] = 3.8 \cdot 10^{-4} M^{0.723},$$

where M is the molecular weight of the polymer and [η] is the intrinsic viscosity of the solution.

Though guar gum powder, when kept in air and moisture tight containers, is stable for over one year, its aqueous dispersion starts decomposing after 24 hours due to bacterial and enzymatic depolymerization. Guar gum is also depolymerized by strong acids, alkalis, oxidizing agents, air (oxygen), heat, and prolonged high shear. Decomposition of the gum is reflected by reduction in its solution viscosity. In an aqueous solution the viscosity of guar gum is also reduced in the presence of salts and hydrophilic liquids (e.g., lower alcohols, acetone, etc.) and hydrophilic solids (sugar) due to their competition for binding of water.

Very fine (>250 mesh) guar gum powder has a tendency to form lumps when dispersed in water. Lump formation results when good stirring is not done during making of its solution. In a formed lump, the powder outside the lump gets wet and surrounds the dry powder inside preventing complete dissolution. However, such fine-mesh powders can be made easily dispersible by granulation, or pH dependent, reversible, and controlled cross-linking with borate or glyoxal (see Chapter 4). Special educting equipment for making paste of fine-mesh guar gum without lumping is also available.

7.27 SPECIFIC INTERACTIONS OF MANNAN CHAIN IN GUAR GUM AND ITS APPLICATIONS[20]

Galactomannans and glucomannans show some specific interactions with other linear polymers in aqueous solutions, polymeric solids, and many small molecules. These interactions arise due to the presence of a *cis*-pair of hydroxyl groups in the C-2, C-3 position of the linear mannan backbone of a these polysaccharides. Some of these interactions and the resulting physicochemical effects are as follows.

Synergistic increase in viscosity with other polysaccharides. Guar gum, when mixed with other linear polysaccharides and proteins, shows a synergistic increase in viscosity, but no gelling takes place. Such synergy has been attributed to the chain–chain interaction of linear guar galactomannan molecules with those of other linear polymers. At the M:G ratio of 2:1 in case of guar gum, there exist only very small nonhairy regions (<8 mannose units long) on its mannan backbone. Interaction of these small, bare blocks of guar galactomannan backbone, with other polymers, can only result in a viscosity increase but no gelling. In these interactions *cis*-hydroxyl pairs have an important role to play, because they can form three-center hydrogen bonds, which are stronger and of longer time duration. Also the hydrogen bonding takes place over extended portions of chain-producing interactions like a zipping of polymeric chains. Hydrogen bonds thus produced have large energy and are durable over an extended period of time.

Interaction with hydrophilic surfaces of solids. Cellulose fibers can strongly adsorb galactomannan on their surfaces, which helps in fiber-to-fiber bonding during paper manufacturing. Such bonding provides an extra overall strength to the paper. There is also an increased recovery of pulp, by aggregation of fines (fibers), which are otherwise lost in the pulp liquor.

Interaction with hydrophilic minerals. In mineral beneficiation of nonferrous metal ores—typically those of copper, nickel, and zinc sulfides—by froth flotation process, the hydrophilic surface of the gangue minerals (the impurities in metallic ores are silica and silicates), adsorb guar gum on their hydrophilic surface and thereby these particles are aggregated into larger flocks and get depressed (sinks). In contrast to this, the mineral particles, which are made hydrophobic by alkyl xanthates, float with the froth. Thus, there is an effective separation of mineral ore from the impurities during a froth flotation process.

Interaction with low molecular weight substances and chiral selective interactions. Enantiomers in certain racemic mixtures show differential and chiral selective interaction in hydrogen bonding to the *cis*-hydroxyl group pair of mannan backbone of guar gum. This was investigated for the first time by the present author. This principle has been used in the development of chiral chromatography for enantiomer separation.

Among the polysaccharides, cellulose- and starch-based materials have long been used as absorbents in column chromatography. Though these polymers are made up of chiral, sugar (glucose) monomers, they do not interact selectively toward enantiomers in a racemic mixture. In contrast to this, guar-galactomannan-based, bead form of column chromatographic media can be made by epichlorohydrin cross-linking, which becomes insoluble in water. Alternatively guar polysaccharide (about 1.0%) can be mixed with TLC-grade silica gel. In both of these cases, chiral selectivity has been observed toward certain racemates. Thus, the racemic mixtures of alpha-amino acids were effectively resolved on silica gel mixed with TLC plates coated with guar galactomannan.

Modulation of carbohydrate and lipid metabolism in the human digestive system. When taken orally, guar gum acts as a soluble, dietary food fiber. It slows the absorption of sugars and lipids from the intestinal track into the blood. This has helped in reducing blood glucose levels in healthy as well as diabetic persons.

By slowing fat absorption, guar gum also controls the blood lipid level, which includes the low-density lipoproteins (LDL), cholesterol, and triglycerides. It has been suggested that synergistic enhancement of the viscosity of food in intestines in the human body and hydrogen bonding play a major role in these beneficial actions of guar gum acting as a food fiber. Additionally the viscous guar gum can entrap digestive enzymes and retard their activity.

Interaction with cross-linking agents. Cross-linking with borate and glyoxal can cause guar gum to form a gel, or at very low concentration, render it water dispersible and slow its hydration. Since the cross-linking is pH dependent it can be reversed and delayed viscosity development can result. Such guar gum formulations have found extensive industrial applications. The mechanism of gelling of guar by borate (pH >7) was discussed earlier. Glyoxal cross-linking, at acidic pH (~3.5), arises due to a bis-acetal formation with two *cis*-hydroxyl pairs on two different chains (Chapter 4). Such reactions are characteristic of polysaccharides containing mannose backbone. These reactions are either not shown by nonmannose polysaccharides or the interaction is very weak.

Cross-linking with transition metal ions. Complexing of transition metal ions (e.g., Zr, Sb, CrO_4, and Ti) by the *cis*-hydroxyl pair on mannan chain results in controlled gelling or viscosity enhancement of guar gum paste. This property of guar gum has been used in enhanced oil recovery and fracturing of an oil well. Metal ion complexed guar is thermally more stable.

According to Dea,[16] these interactions arise due to hydrogen bonding associations, which are of a higher energy and of a longer time duration when compared to normal hydrogen bonds. In some of the oil-field applications, metal-complexed guar galactomannan and its derivatives are more effective than most other polysaccharides.

7.28   USES AND APPLICATIONS OF GUAR GUM

Guar gum was initially developed as a substitute for LBG in the U.S. paper industry. After its industrial production commenced in the late 1950s, many other applications for guar gum were found. These are based on the functional properties of this hydrocolloid as a viscosifier, stabilizer, emulsifier, and so forth. Being cheaper among most hydrocolloids and readily available, it has found many new applications and replaced many earlier used gums in some of the specific applications. Derivatized guar gum products have further enhanced the scope of their wider applications. More important applications of guar gum are briefly discussed next.

Oil- and gas-well drilling and fracturing. In oil-well drilling, guar gum acts to prevent water loss from the viscous drilling mud and as a suspending agent for bentonite clay used in the mud. These functions are performed very satisfactorily by guar gum and it is more cost effective than most of the other mud thickeners. It has the limitation of being thermally less stable compared to xanthan gum at temperatures >100°C. To a large extent, this limitation has been overcome by employing a guar-hydroxypropyl derivative, which is thermally more stable. For enhanced oil recovery or oil-well stimulation by fracturing of the oil well, a popping agent (e.g., sand suspended in thickened guar gum or a hydroxypropyl-guar gum solution) is pumped into the well under pressure to pop up and widens the cracks in the rock formations. This permits more oil and gas to percolate into the well. Frequently cross-linking agents (e.g., borate or transition metal ions [zirconium and titanium]) are added to cause in situ gelling of injected guar gum paste. Having completed the fracturing, the gel is broken (by enzymes or acid) and flushed out, leaving minimum residue after the break.

Paper industry. Guar gum was initially introduced as a substitute for LBG in the paper industry, where it worked well in improving recovery of pulp and providing additional strength to the paper; this application still continues. During the past several years, the use of guar in paper making has been reduced in favor of cationic starch, which also reduces the cost, for cheaper grades of paper.

Explosives. For making water-resistant slurry and gel explosives based on ammonium nitrate, a gel of the latter is made in guar gum paste. Guar gum prevents leaching out of the explosive material and such gelled explosives are suitable in the mining industry where they can be given any desired shape. Water gels or slurry-based explosives were first introduced in 1958. These were mixtures of ammonium nitrate, TNT, aluminum or magnesium powder, water, and a gelatinizing agent, which usually was guar gum and a cross-linking agent such as borax. (Cross-linking is due to covalent or hydrogen bonding.) Later, aluminum powder and other metallic fuels were also incorporated into the gel explosives. Guar proved to be a vastly better gelatinizing agent for these gel explosives. In addition nonexplosive sensitizers were developed that could replace the TNT if desired. When the highest possible concentration of strength is needed, large quantities of TNT are still used. Water gels have many advantages over conventional solid explosives.

Mining industry. Mining is one of the major industries using guar gum for effective beneficiation of nonferrous metallic ores by the froth floatation process. During concentration of ores by froth flotation, the metallic ores are rendered hydrophobic by interaction with a floatation agent (e.g., alkyl xanthates) and these float with the froth. The impurities or the gangue mineral (silica, silicates) adsorb guar gum on their surface to become hydrophilic and these are depressed. Guar gum has proven to be an economical and effective depressant compared to most other hydrocolloids. Its major use as a depressant is in nonferrous ores of copper, nickel, gold, uranium, and potash minerals.

Tobacco industry. The tobacco industry uses guar gum for moisture retention in tobacco and for making reconstituted tobacco preparations from tobacco leaves, which are broken or powdered during processing. Generally guar gum acts as a better moisture-retention agent than other gums.

Cosmetics and pharmaceutical industry. Cosmetics and pharmaceuticals are the new emerging industries to use guar gum products. Highly refined, modified, and derivatized guar gum products are used by these industries. The gum products have functions of thickening, suspending, binding, and emulsifying in cosmetic products like shampoo, conditioner, moisturizer, and toothpaste, and in many liquid medicinal preparations. Cationic and hydroxypropylated guar gum are extensively used in cosmetics. In medicinal tablets, guar gum is used as a binder and drug-release control agent. Depolymerized guar gum has been found to be a good bulking agent and a source of fiber for dietetic food. As a food fiber it helps in sugar and lipid metabolic control, particularly for diabetic and heart patients.

Textile printing and dyeing. In India and in many other countries, one of the major applications of guar gum has been in textile printing, particularly for printing of cotton with fiber-reactive dyes. Depolymerized and anionic derivatives of guar gum, particularly the carboxymethyl (CM) and sulfonic derivatives of guar gum, have proven to be good and reasonably cheaper substitutes of sodium alginate as a print paste thickener. For polyester printing using disperse dyes, the use of guar gum has now been largely replaced by thickeners based on tamarind kernel powder (TKP), which is cheaper.

TKP is often used for disperses dyes, but use of guar gum continues in cotton printing. TKP is frequently blended with guar gum to reduce the cost of a print paste thickener. A large amount of guar-gum-based products are exported from India for use in carpet printing and dyeing, where it works very well. In dyeing it acts as an antimigrant. Guar gum is also used in blends of textile size where it improves film forming and spreading of other sizing materials (e.g., starch).

Food industry. Like most of the food hydrocolloids, guar gum has the functional properties of modifying a water system by increasing its viscosity and acting as a binding, stabilizing agent for suspensions and emulsification agents. These are the common requirements of any food hydrocolloid additive. Hence, guar gum is used as an economical additive in many food preparations. It has been used as a stabilizer and water crystallization inhibitor in frozen foods and ice creams. In soups, sauces, and ketchup guar gum is used as a thickener and stabilizer. In many baked foods, it prevents staleness, reduces crumb formation, increases the shelf life of a product, increases the dough volume, and prevents loss of moisture. It is also used in soft cheese, noodles, icings, dressings, meat bindings, and spreads. In beverages and instant drinks it improves mouthfeel and aftertaste. It is used as a binder in pet food. Depolymerized guar gum is a consumer-acceptable food fiber, which has been used in preparation of low calorie food. About 40% of the total guar gum produced is used as a food additive. Use of guar gum in food in India is still low, but it is fast increasing with the growth of the food industry.

Other applications. Guar gum has found many other minor uses, for example, in wild firefighting, where it keeps the moisture and fire retardant substances at the targeted site. At very low concentration it reduces the friction due to turbulent flow of water through a pipe and helps in firefighting by increasing the flow rate of water through a pipe. This principle is also used in sprinkler irrigation to increase flow rate of water. It is added to agriculture spray formulations, where it makes the active ingredients stick to the plants and makes them more effective. In the building construction industry, guar gum has been used in cement-mortar formulations, as a compacting agent, a water retention aid, and to improve the cement curing process. This results in improving the ultimate strength in construction.

7.29 CONCLUSION

Currently there are more than 30 natural and modified polysaccharide-based hydrocolloids (gums) in commercial production and use. A hydrocolloid should be produced in a quantity sufficient to make it economically viable. Guar gum occupies one of the top positions among industrial polysaccharide gums. Many of the functional properties of different gums can be similar, which make their applications and uses interchangeably possible. When the price difference of these products is large, the cost effectiveness is an important factor in making the right choice. Alternate products can be used in some applications, but there are other, more specific applications in which an alternate choice is limited due to a particular functional property. Thus, in the case of gum arabic, good emulsification is the main functional requirement of the hydrocolloid, whereas for pectin, gelling is the main functional requirement. These requirements may not be met satisfactorily by an alternate, cheaper gum substitute.

To an extent, the functional properties of a hydrocolloid can be altered by physical or by chemical modifications. Plant scientists can also work on genetic modifications of alternate and abundant plants to incorporate any desired functional behavior into their gums.

In India, guar gum manufacturers should work to develop value-added products based on guar gum and take maximum advantage of them.

REFERENCES

1. Hymowitz, T., The trans-domestication concept as applied to guar, Econ. Bot., 26 (1972): 49.
2. Whistler, R. L. and Hymowitz, T., Guar: Agronomy, Production, Industrial Use and Nutrition, Purdue University Press, West Lafayette, IN, 1979.
3. Loock, E. E. M., The carob or locust tree, Ceratonia siliqua, J. S. Afric. Forest. Ass., 4 (1940): 78.
4. Rowland, R. L., The use of guar in paper manufacture, Chemurg. Dig., 4 (1945): 369.
5. Anonymous, Tropical legumes: Resources for the future, National Academy of Sciences, Washington, D.C. (1979): 09.
6. Paroda, R. S. and Arora, S. K., Guar: Its Improvement and Management, The Indian Society of Forage Research, Hissar, India, 1978.
7. Whistler, R. L., Chem. Ind. 62 (1948): 60.
8. Anderson E., Endosperm mucilage's of legumes: Occurrence and composition, Ind. Eng. Chem., 41 (1949): 87.
9. Kapoor, V. P., personal communication, National Botanical Research Institute, Lucknow, India, 2004.

10. Whistler, R. L. and BeMiller, J. N., Eds., Industrial Gums, 2nd ed., Academic Press, New York, 1973, 303; 3rd ed., 1993, 181–225.

11. Dalal Consultants and Engineers, Study on Guar Crop, a survey carried out for the Agricultural and Processed Food Products Export Development Authority, New Delhi.

12. Seaman, J. K., In Handbook of Water Soluble Polymers and Resins, R. L. Davidson ed., McGraw-Hill, New York, 1960, chap. 6.

13. Dhugga, K. S., Plant cell wall polysaccharide biosynthesis: Industrial applications, Seminar at the Pioneer Hi-Bred International, Johnston, Iowa.

14. Mathur, N. K., unpublished, 2002.

15. Heyne, E., and Whistler, R. L., J. Am. Chem. Soc., 70 (1948): 2249.

16. Dea, I. C. M., Structure/function relationships of galactomannans and food grade cellulosics, In Gums and Stabilizers for Food Industry 5, G.O. Phillips, P.A. Williams and D.J. Wedlock, eds., RL Press, Oxford, 1989, 373.

17. Couch, R., Creger, C. R. and Bakshi, Y. K., Proc. Soc. Exp. Biol. Med. 123 (1966): 263.

18. Ahmed, Z. F. and Whistler, R. L., J. Am. Chem. Soc., 72 (1950): 2524.

19. Baker, C. W., Carbohydr. Res. 45 (1975): 237.

20. Dea, I. C. M. and Morrison, A., Chemistry and interactions of seed galactomannans, Adv. Chem. Biochem. Carbohy., 31 (1971): 241.

8 Locust Bean Gum, or Carob Gum

8.1 HISTORICAL INTRODUCTION[1]

The coast of the Mediterranean Sea consists of dry mountainous steep slopes and is semiarid, where normal plowing of the land is not possible. As a consequence, common agricultural crops do not grow there. Such regions are only suitable for tree crops, which have often yielded valuable harvests in this region. One such tree crop is that of the carob tree, also called the locust bean tree (*Ceratonia siliqua*), which holds special promise in the hot, coastal Mediterranean climate, which has an extended dry season. Carob tree is noted for its drought resistance, particularly in a region where rains are uncertain and irrigation is not possible. For centuries this legume tree has contributed to the economy of the Mediterranean basin.

Perennial and evergreen carob tree grows in the Mediterranean region of Southern Europe, Northern Africa, and some adjoining areas of the Middle East. The carob tree was brought to these regions from the Egypt of Pharoahnian periods. The Palestinians, Arabs, Greeks, and Romans used carob seed gum for binding of the mummies of their dead nobles. The climate of this new habitat for the carob tree in Europe was found very suitable for its growth, where the tree flourished. Originally indigenous to the Eastern Mediterranean region only, the carob tree now grows and it is cultivated in many other regions of Europe and Africa. Carob is one of the several legume plants, which produce galactomannan-containing mucilage, and have been known, used, and cultivated for many centuries.

The name *carob* has its origin from the Hebrew word *kharuv* meaning "saber." In Arabic, this tree is called *al kharoubah*; while in French, Italian, and Spanish, the names for the carob tree are *katoube*, *carruba*, and *algarroba*, respectively. The term *carat* has been derived from the legend that the Arabs used carob seed for weighing precious metals (gold and silver) and stones.

Until recently, the carob tree was not grown elsewhere in the world. Millions of hectares of semiarid regions, similar to the costal Mediterranean, are present in the subtropics, which appear to be well suited to grow the carob tree. The reason for lack of carob plantation elsewhere may be because it is a very slow-growing tree. It takes 10 to 12 years before a carob tree starts bearing fruits, which is the source of carob gum. Some roadside plantation of carob trees has been successfully carried out in California, though as yet there is no report of their commercial exploitation for the extraction of locust bean gum, which is a galactomannan polysaccharide.

One of the commercial products obtained from carob tree is its seed endosperm polysaccharide, which is a unique galactomannan gum commonly known as carob gum, locust bean gum or simply LBG. LBG is the refined endosperm powder from carob seed.

Why the plantation of the carob tree, which is the only tree producing a unique galactomannan has not been undertaken elsewhere? I think the answer could be that in this era of science and technology, the possibility exists of developing an alternate, cheaper, and better substitute for LBG as another unique commercial hydrocolloid. This may be possible by modification of a cheaper and abundant hydrocolloid (e.g., guar gum) or from some still unexplored plant as a source of gum. This might be the reason for not undertaking large-scale, new plantations of carob trees elsewhere.

In contrast to the perennial trees as sources of galactomannans, priority is now being given to annual legume crops as their source. Though not much success has so far been gained to find an equivalent or better substitute for LBG,[2] more plants are now being explored as sources of galactomannan gums.

8.2   CAROB TREE

The carob tree belongs to the pea family, of the larger Leguminosea family. Its subfamily is Fabaceae and Casalpiniaceae.[3,4] The carob tree can be propagated by the seed as well as by its cuttings. The tree grows to a height of about 9 to 15 meters and it has luxuriant and perennial foliage. This handsome tree has pinnately compounded (feather formed), glossy evergreen leaves with thick foliage. After 10 to 12 years of maturing, the tree starts flowering, which is followed by bearing of flat, leathery pods that are 10–20 cm long, 2–5 cm wide, and 0.5–1.0 cm thick. When fully ripened, the pods turn brown, and finally black when dry.

The carob tree, which has multiple uses, was known in Middle East countries even before the Christian era. Arabs used carob fruit as a feed for their horses and milk-producing cattle, while it served as a food for poor people. In the Bible, carob seed finds reference as a constituent of St. John's bread. During the times of Pharaohs, carob wood was extensively used in Egypt and Palestine for making furniture, idols, and for buildings the temples of worship. A paste of locust bean gum was used in Egypt for mummifying the dead. In Greek and Roman books of pharmacy, Dioscorides, the Greek man of medicine, referred to the curative properties of carob fruit. These medicinal applications included laxative properties and many other remedial cures. Modern industrial uses of LBG and its use as a food hydrocolloid have now been done for nearly a century. For its first nonfood application, it was adopted by the U.S. paper industry as an additive to wood pulp in which it increased the overall strength of paper.[5]

Carob pods make an excellent fodder for livestock. Ripened pods, when not collected from the tree, begin to fall on the ground. Each pod contains 5–15 large, hard, brown seeds that are embedded in a sweet, edible pulp. Carob pulp contains a higher percentage of sugar than sugar cane or sugar beet. The pulp of the fruit is edible and it has been used as a substitute for chocolate flavor in confectionary items, such as bread, cake, candy, and breakfast cereals. The pulp is also fermented to make it into wine.

In Israel alone, there are over 250,000 carob trees in the Levant (Eastern Mediterranean region) hillside forest area. In Cyprus, carob trees are planted to a density of about 90 trees per hectare. Depending upon the weather conditions, the trees can annually yield 2 to 4 tons of carob pods per hectare. The yield of carob pods is highly variable from place to place. Some large and old carob trees have been reported to yield up to a ton of the pods in one season. Cyprus is one of the major carob-producing countries. It exports over 45,000 tons of deseeded carob pods annually to Europe, where its seeds are processed for its valued galactomannan polysaccharide, LBG.

Prior to World War II, carob seed or the kernel was also exported to the United States where its gum was extracted and used in the paper industry. During World War II (1939–1946), the supply of LBG from the Mediterranean countries was greatly reduced. It was during that time that a search for an alternate and suitable galactomannan as a substitute for LBG was undertaken by U.S. paper companies. Ultimately, the guar plant emerged as a rival to substitute for LBG in many of its applications.

8.3 CAROB SEED, THE TRADITIONAL SOURCE OF A GALACTOMANNAN GUM[2,4]

The traditional and earliest known galactomannan polysaccharide is the locust bean gum, which is obtained from the carob pod (Figure 8.1).

Some carob trees automatically shed the ripened pods, whereas on many other trees the pods have to be harvested by shaking the branches of the tree. In either case, the pods are collected manually on the ground. Morphological study of the carob endosperm has shown that the cell structure of different layers of seed cotyledon is not as variable as in the case of guar seed endosperm. Still a clear distinction in the composition of LBG powder can be made from the difference in water solubility of fractions of the gum obtained from carob seed. The gum fractions, having different solubility, differ in their mannose-to-galactose ratios (M:G).

The carob seed coat is dark-chocolate in color. This dicotyledonous seed contains a hard yellow embryo or the germ, which contains the genetic material of the plant, and it is rich in proteins. Corneous-type large layers of white, translucent endosperm surround the germ, which is the main and desired polysaccharide product in the processing of carob seed. A tenacious, dark brown husk covers this endosperm.

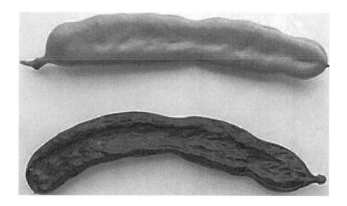

FIGURE 8.1   Carob pods.

Locust bean gum is the second most important galactomannan after guar gum and its annual worldwide production stands, next only to guar gum. The difference between the two gums is their plant source. Whereas LBG is derived from a perennial tree (carob), guar gum is obtained from an annual crop (guar, *Cyamopsis tetragonolobus*). Many of the industrial uses of LBG have now been restricted due to economical considerations in favor of guar gum. LBG is also known as *carubin*, and it is approved as a safe food additive by the European Union. It has been assigned the EU food additive number E410.

8.4   CAROB GUM OR LOCUST BEAN GUM[4]

Upon ripening in the months of October to November, the green color of the pod becomes dark brown. The ripened pods are easily detached from the tree when shaken. The ripened pods can also be shaken off the tree with poles and these are then collected manually on the ground. On an average a tree produces 200–250 Kg of carob pods in one season.

Some gigantic trees have been reported to bear up to a ton of pods annually. The best LBG is produced from seeds harvested in Sicily. Some of the trees in Sicily are supposed to have been planted in the 16th–17th century. The world harvest of locust bean kernel is estimated to be between 35,000 and 38,000 tons.

LBG or the carob gum, which is derived from carob kernel, is basically a galactomannan polysaccharide. The carob pod consists of approximately 90% pulp, 8% kernel, and 2% other constituents. After sun-drying the pods are cleaned and kibbled mechanically to separate the pulp and the kernel. The composition percentages of the various components of the carob kernel are shown in Table 8.1.

8.5   AGRONOMY OF CAROB TREE[2]

The Mediterranean seashores of steep and rocky slopes, which are unsuitable to grow annual agricultural crops, have proven to be suitable for the growth of carob, olive, and certain other trees. These trees yield a valuable harvest of marketable products for this semiarid region.

Compared to many fruit-bearing trees, carob is less demanding on the soil and grows well on rocky hillsides. Its roots can penetrate deep into the crevices of the rocks and thrive on little water, but it cannot withstand water logging. In most places, carob is planted along with other trees of the same regional origin, for example, the olive tree. Being a nitrogen-fixing legume tree, it helps in improving the soil and promotes forestation. Carob being a hardy tree also helps in preventing soil

TABLE 8.1
Composition of Carob Seed

Husk (Cellulosics and pigments)	25%–30%
Endosperm (Galactomannan)	40%–45%
Germ (Protein and nucleic acids)	20%–30%

erosion. Because of its spreading crown, it is regarded as an indispensable ornamental and shade tree. With continuing urban developments in the carob-growing areas in Europe and Africa, a long-term projection for LBG as an industrial commodity sounds as decreasing production and increasing price of the gum.

In the United States, considering all the regional and climatic factors, carob trees have been planted in the hilly, residential areas of California, mainly as shade trees. As of yet these plantations have not been used for harvesting gum. One major limitation of the carob tree is that it matures very slowly and the fruit yield increases with the age of the tree. Only after 12 to 15 years when it is fully grown, the tree starts bearing fruits, after which it continues to bear fruits for up to 50 years or more. Some carob trees can have a life span of over 100 years. Overall this tree grows well where annual rain averages 400–550 mm. Carob can survive in a temperature range of 5°C to 40°C, but it does not tolerate subzero temperatures.

Depending on the area of plantation, the tree starts flowering in the months of January and February. It bears fruits from April to May, which fully mature by September–November. World crop of matured carob pods amounts to nearly 50,000 tons.

For planting of new trees, a pretreatment of carob seed by dilute sulfuric acid helps in early germination of the plant. Male carob trees only produce pollens, while female trees bear pods. Frequent wetting of tree leaves due to dew during nights throughout the year, can cause leaf disease in carob, which is due to a red scale fungus. During humid summer rainfall, a powdery mildew fungus can infect the fruits.

8.6   REGIONAL CONSIDERATIONS[4]

The carob tree is mainly grown in Southern Europe's Mediterranean countries—Spain, Portugal, Italy, Greece, and Turkey—and the Mediterranean islands of Sardinia, Sicily, Crete, Malta, and Cyprus. In the North African Mediterranean region it grows in Morocco, North Algeria, and Libya. It also grows in the adjoining areas of the Middle East—Egypt, Syria, Lebanon, and Israel—where some very old and surviving carob trees are thought to date back to the time of Christ.

Trial planting of the carob tree has been done in Australia, Argentina, United States, Mexico, Malawi, Rhodesia, and South Africa, but as yet none of these countries have been identified as a producer of LBG.

Labor cost for picking and sorting of carob pods, which is done manually, determines the economy of the product gum. Currently there is much variation in labor cost in African, Middle Eastern, and European countries. Hence, the production cost of LBG in Europe is high, making it desirable to find a cheaper substitute.

During and just after World War II, several other uses of LBG as an industrial hydrocolloid were discovered. LBG was used in thickened aqueous systems as a rheology control and modifying agent for example, in food, textile printing paste, slurry explosives, and drilling mud thickener in the petroleum oilfields, to name a few. The history of industrial production of carob seed gum is nearly eight decades old. Two English companies (Tragasol of Hooton and Ellis Jones of Stockport) and an Italian company (Cesalpinia in Milan) are currently doing industrial production of LBG. The names and the ownership of these companies have since then changed, but they continue to work with a variety of plant hydrocolloids, including LBG.

Carob seed galactomannan or LBG is currently used the world over as a hydrocolloid for rheology control and related applications in food and other industries. LBG is marketed under various trade names, including Carob Gum, Gum Gatto, Gum Hevo, Jandagum, Lakee Gum, Rubigum, Lupogum, Lupsol, Gum Tragon, Tragarab, and Tragsol.

8.7 IMPORTANT CAROB PLANTATION REGIONS IN THE WORLD

Carob plantations are in the following Mediterranean countries:

> Spain—Catalonia (Tarragona), Valencia (Castellon), Murcia (Cartagena), Andaulsia (Malaga), and Balearic Island (Majorca, Minorca, and Ibiza).
> Italy—Sicily, Apulia, Calabria, and Sardinia
> Greece—Crete, Peloponnesus, and Phthiotis
> Cyprus—The whole island
> Portugal—Algarve
> Turkey—Izmir and Mersin
> Israel—Most of the area near Jerusalem
> Algeria—Oran, Bougie, Constantine, and Algiers
> Morocco—Casablanca, Essaouira, and Agadir

In addition to these traditional carob-growing areas, some carob plantations have also been done in Australia and the United States.

The best locust bean gum comes from Sicily, and some of the trees there were planted centuries back.

8.8 MANUFACTURING OF LOCUST BEAN GUM[6,7]

Most of the LBG from the carob seed is processed in Europe, but some of it is also manufactured in African countries, particularly in Morocco.

Locust bean gum, which is marketed world over, is classified according to its purity and designated as

1. Normal high grade
2. Industrial grade
3. Technical grade

These three types of products are of decreasing purity and viscosity, and these are generally available as 100–150 mesh powders. The gum powders may contain variable amounts of impurities from the germ (proteins) and the husk (insoluble fibers) of the seed, but these impurities do not render LBG of any of these grades nonedible.

The presence of the germ portion (proteins) as an impurity renders LBG paste more prone to fermentation because of the enzymes present therein.

Production of high-grade LBG is much more difficult when compared to that of guar gum. The company Cesalpinia of Italy has invented an improved method of making LBG powder. When dehusking of seed endosperm is not properly done, the product gum contains brown specks of the husk, which renders it brownish in color. In an improved and alternate procedure, better dehusking is achieved by heating of the seed to a temperature of 120°C–150°C for ~45 seconds in a tilted, rotating kiln. This is followed be milling in a series of roller mills when the husk and the germ are removed simultaneously.

Dehusked seeds are longitudinally split into endosperm halves, while differential grinding and sieving separates the endosperm from the germ. Endosperm halves are again crushed and then coarsely ground in machines. Finally the endosperm is ground into fine (100–150 mesh) powder in a hammer mill. Since traces of germ portion containing hydrolytic enzymes cause depolymerization of LBG in solution, removal of germ from the LBG powder is carried out effectively.

Like most of the polysaccharide gum solutions, those of LBG when stored for long periods, that is, 24 hours or more at room temperature are subjected to bacterial breakdown, and hence the addition of proper preservatives to the gum powder is recommended. Benzoates and sorbates are preferred as preservatives when LBG is meant for food applications. Many more preservatives (e.g., formaldehyde and chlorinated phenols) are used in nonfood applications. Compared to guar gum, heat and shear depolymerization of LBG during its processing is less.

8.9 FLOW SHEET SHOWING VARIOUS STEPS OF MANUFACTURING LBG

The following flow sheet represents various steps in the manufacturing of LBG.

Matured, dried, and cleaned carob pod (starting material)
↓
Kibbled to separate kernel and pulp → Pulp is separated
↓
Kernel or seed is now separated from pulp
↓
Seeds heated in a rotary kiln at 120–150°C for ~45 seconds
↓
Dehusking of seeds by milling and sifting → Husk and germ are removed
↓
Purified and dehusked endosperm or the splits are obtained
↓
Endosperm crushed into smaller pieces
↓
Hammer milling
↓
Produces 100–200 mesh LBG powder, which is sieved according to the particle size
↓
Testing, quality control, and packing for marketing

8.10   POTENTIAL AMOUNTS OF MARKETABLE LBG AND PRICE VARIATION[7]

Currently 15,000–20,000 tons of LBG is marketed worldwide. Being a product from a perennial tree, there is far less variation in annual production of LBG compared to that of guar gum, which is obtained from an annual crop. At the same time the annual production of LBG cannot be increased as and when it is desired. Though the carob tree grows well in hot weather, frequent subzero temperatures and long wet weather can reduce its seed-crop yield, seed size, and its gum content. These factors can ultimately result in lowering of the annual production of LBG.

Considering the fact that total production of LBG is likely to remain stagnant at the present level, several attempts have been made to find substitutes from other trees (e.g., tara shrub) or to modify the functional properties of abundant guar gum to replace it for LBG. Such attempts have met with very limited success.

8.11   FUNCTIONAL PROPERTIES OF LBG[3,8,9]

In LBG molecules, the presence of continuously substituted large blocks of galactose grafts on the mannan chain, which is separated by blocks of bare backbone, has resulted in certain unique and specific functional properties. Such properties are shared to a far lesser extent by other galactomannans, including tara gum (M:G = 3:1) and even those gums that have an M:G ratio very close to or even lower than LBG. It is this unique structural feature of LBG that makes it indispensable for certain applications in the food industry, which will be discussed later.

This peculiarity in the distribution of galactose grafts on the mannose backbone of LBG—that is the occurrence of large blocks of substituted portions (called hairy region) of as many as 25 or more mannose units separated by even larger, unsubstituted portions (called the nonhairy or bare portions) of the backbone—has conferred certain unique functional properties to LBG. Though the average M:G ratio in LBG is 4:1 (more exactly it is 3.5:1), this gum can be separated into three main fractions:

1. 35% cold-water soluble (M:G = 1.2:1)
2. 52% hot-water soluble (M:G = 3.3:1)
3. 10% insoluble (M:G = 5.2:1)

There are many other intermediate fractions containing continuously decreasing percentages of galactose. This indicates the heteropolymolecular and polydisperse nature of LBG. Lower galactose content of LBG in some fractions increases the flexibility of its molecule when compared to those of guar or fenugreek gums.

From the viscosity measurement of industrial samples of LBG at 1% aqueous concentration, which comes to 2500–3500 cps, an average molecular weight of approximately 50,000 dalton has been suggested for commercially sold LBG. A more recent estimate of the original and undegraded lab sample of LBG polysaccharide puts the molecular weight in the range 300,000–350,000 dalton. This range of molecular

TABLE 8.2

Comparative Compositions of Typical

Commercial Samples of Guar Gum and LBG

Component	Guar Gum (%)	LBG (%)
Galactomannan	75–85	70–80
Moisture	8–12	10–15
Protein	3–6	5–10
Insoluble fiber (cellulose)	2–3	5–8
Ash (minerals)	0.5–1.0	1–2

weight corresponds to the presence of ~2000 monomer units (mannose + galactose) in a galactomannan molecule.

In common with other galactomannans, and by virtue of the presence *cis*-hydroxyl groups in the constituent sugars, LBG can undergo general interactions of galactomannans, as discussed in Chapter 3. Thus, LBG can be gelled by borax at alkaline pH (>8) and it also forms complexes with certain transition metal ions.[10]

Commercial processing of carob seed to extract the gum is more difficult as compared to that of guar seed, and it does not yield a product gum that is as pure as commercial guar gum. For comparison, typical compositions of commercial samples of the tree legumes (carob) galactomannan gum and guar gum are show in Table 8.2. It is observed that the purity in terms of galactomannan polysaccharide in LBG is not as high as that of guar gum.

It was earlier mentioned that LBG is produced in three different purity grades. These are classified according to their galactomannan content. According to an Italian manufacturer, the purest grade of gum has a viscosity of 3500–4000 cps at 1% concentration in water, as measured by a Brookfield viscometer, spindle No. 3 at 20 rpm and 25°C. Further, this quality of gum is made in different mesh sizes of 150, 175, 200, and 250 standard U.S. mesh. High quality and pure gum powder is almost colorless and free from specks of brown husk. Galactomannan content of such a high purity product is 80%–90% on a moisture-free basis.

Commercial LBG contains about 80% galactomannan, 4% pentosans, 6% protein, 1% insoluble fiber (cellulose), 1% ash, and 1% ether extractable (lipids) on a dry basis (Table 8.3). LBG samples contain 5%–10% moisture. Compared to high-grade LBG (viscosity of 3500–4000 cps), technical-grade LBG products have a viscosity range of only 800 to 1000 cps. For viscosity measurement, a 1.0% solution of LBG is made in water, which is preheated to 90°C–100°C (10 minutes), and then the solution is cooled to 25°C. The viscosity measurement is finally done using a Brookfield viscometer. Unlike guar gum, fine-grade LBG, which is a white powder, is soluble only in hot water, developing full viscosity on heating at 90°C for 10 minutes.

The solubility of any β(1-4)-linked linear glycan (e.g., galactomannans) depends on the frequency and uniformity of grafts on the polymer backbone as well as on their distribution on the mannan chain. When large blocks of bare backbone are present in the molecules of a polysaccharide, they can interact strongly, forming multiple interchain hydrogen bonds or time-durable junction

TABLE 8.3

Composition of Commercial Grade LBG

Component	High Grade (%)	Medium Grade (%)	Technical Grade (%)
Galactomannan	80–88	75–80	72–78
Moisture	4	6–8	10–12
Pentosan	3	4–5	4–5
Protein	2–3	6–8	8–10
Insoluble fiber	1	2	2–3
Lipids	1	1–2	2
Ash	1	1–2	1–2
Viscosity 1% solution	3500–4000	800–1000	500–800

zones. This can render the polysaccharide insoluble in cold water, which is the case with LBG.

This polysaccharide exhibits a semigel or a thixotropic rheology, which is typical of partially soluble hydrocolloids such as LBG. Since some of the hydrogen bonding interactions between strands of molecules break when heating, a part of LBG dissolves in water and a part of it remains insoluble. The insoluble gum part is suspended into the form of a pseudoplastic solution.

Like most plant hydrocolloids, solutions of LBG are shear thinning or pseudoplastic in nature.[3] LBG can show thixotropic and semigel rheology at higher concentrations. Viscosity of a LBG solution decreases with an increase in temperature and shear rate. After adding a water miscible organic solvent (e.g., an alcohol) to a solution of LBG, or at a sugar concentration of ~ 60%, it forms a weak gel. In contrast to this, precipitation takes place in the case of guar gum.[4,5]

Locust bean gum retards ice crystal formation due to structured gel produced at the solid–liquid interface. This particularly occurs on freeze–thaw cycling, which encourages the frustrated crystallization of the galactomannan and causing it to form a gel.

LBG promotes phase separation with skimmed milk powder showing synergistic viscosity with casein and becoming slightly thixotropic. It forms a biphasic system containing casein micelles within a continuous polysaccharide network. LBG may be usefully combined with xanthan polysaccharide, with which it shows viscosity synergy, and with carrageenan polysaccharide. It has been suggested that LBG is adsorbed onto the superhelices of carrageenan to strengthen a three-dimensional structure, resulting in gelling.

Like most other polysaccharides, LBG solutions are subjected to bacterial breakdown and the use of appropriate preservatives in its samples is recommended. Compared to guar gum, heat and shear depolymerization of LBG is far less.

Though many patents have been granted on LBG derivatization, these derivatives are not produced on an industrial scale. This is because of the initial high cost of the gum, which is the starting raw material for making its derivatives. LBG can be rendered cold-water soluble by derivatization, for example, carboxymethylation, phosphorylation, or hydroxalkylation. It can also be made cold-water soluble by blending it with sugar or guar gum.

The following equation has been suggested as a correlation between molecular weight and intrinsic viscosity of LBG solutions,[11]

$$[\eta] = 9.28 \times 10^{-6}. \, M^{0.98}$$

where M is the molecular weight of LBG.

8.12   RHEOLOGY OF LBG

Fine-grade LBG is a white powder that is only partially soluble even in hot water. It develops full viscosity only when heating to ~90°C for 10 minutes. As mentioned earlier, the solubility of any linear β-(1→4)-linked glycan depends on the frequency and uniformity of grafts on its linear polymer backbone as well as on the distribution of these grafts on the mannan chain. Partial solubility of LBG can give rise to a desirable, mixed sol-gel or the thixotropic rheology to a hydrocolloid, as is the case with LBG. Since the hydrogen bonding interaction breaks to some extent on heating, a part of the polysaccharide dissolves into a viscous, pseudoplastic solution in which some undissolved part of the polysaccharide remains suspended to give it a translucent appearance to its solution.

Highly pure LBG has a high viscosity of 3500 to 4000 cps. The commercially marketed carob gums have a much lower viscosity in the range of 800 to 1000 cps.

Some chemically modified LBG products have been made fully water soluble by introducing additional grafts on mannan backbone. For example, the anionic, carboxymethyl groups, when introduced into LBG by derivatization reaction using monochloroacetic acid, gives a soluble product. Binary gelling properties are retained in these soluble derivatives.

8.13   MECHANISM OF BINARY GEL FORMATION WITH LBG[12]

Locust bean gum differs from guar gum in that it forms weak, thermally irreversible gel by association of galactose-free backbone regions and it has poor freeze–thaw properties. Galactose grafts, present on the mannan backbone of any galactomannan, reduce strong chain–chain interactions to a certain extent. Such interactions, which can normally result in junction zone formation between two chains of different molecules, can still take place when 10 or more unsubstituted mannose monomers on the polymer backbone are present in a continuous row. Such chain–chain interaction produces nanocrystalline regions, which dissociate in hot water, resulting in solubilization of LBG. There is also a strong interaction between unsubstituted portions of LBG backbone with certain other polysaccharide molecules, typically those of carrageenan, agar, and xanthan. Thus, these nongelling, or weakly gelling, polysaccharides form more firm, binary gels in the presence of small amounts of LBG, compared to those formed by these polysaccharides when present alone. These polysaccharides (carrageenan, agar, and xanthan) can also undergo gel formation at otherwise nongelling concentrations when they are mixed with LBG. For economical considerations, formation of such binary gels has found industrial applications for making gelled food. It has been suggested that even at nongelling concentrations, carrageenan, xanthan, and agar have partially organized (tertiary or the helical) structures. Such partially organized structures exist even in solution at room temperature.

When the temperature is lowered below room temperature, the molecular chains of the polysaccharides reorganize further into a three-dimensional structure. This arises due to the formation of time-durable junction zones between polymer chains, which result in gelling.

Addition of LBG at a low concentration further reinforces the junction zones by interaction of the helical structure of partially gelling polysaccharide with the long and bare regions of the mannan backbone of LBG. One LBG molecular chain can interact with several polysaccharide helices to reinforce and stabilize a three-dimensional network, with liquid imbibed into the unorganized portions of the structure. Such a highly cross-linked structure can hydrate and immobilize the viscous fluid into a gel.

8.14 SPECIFIC STRUCTURAL FEATURES OF LBG[13]

Hirst and Jones (1948), and Smith (1948) were the first to carry out chemical structural studies of LBG. They carried out exhaustive methylation of purified LBG polysaccharide. This was followed by hydrolysis of the fully methylated LBG and identification of the methylated mono- and oligosaccharides produced on hydrolysis. From these studies, the chemical structure of LBG and including the mode of linkage of various sugar units (mannose, galactose) was deduced.

As a general method to study the structure of any polysaccharide, the structure of LBG can also be deduced from the structure of the methylated D-mannopyranose and D-galactopyranose and from those of the methylated oligosaccharides produced by partial hydrolysis.

There has been some variation in M:G ratios in samples of LBG from different geographical regions. Average percentage of mannose sugar in LBG was found to vary in the range 73% to 89%, the rest (27% to 11%) being galactose. It has been reported by different workers that such variations can also be attributed to a varying degree of purity of samples that were used in the study and also due to the variation of the samples from plant sources from different regions.

Locust bean galactomannan is structurally similar to guar gum. Its molecules consist of a linear β(1→4)-linked D-mannopyranose backbone with branch points from their 6-positions linked to D-galactopyranose by an α(1→6) link. There are, on an average 3.5 (varying from 2.8 to 4.9) mannose residues in the polymer backbone, for every grafted galactose residue.

Locust bean gum is thus a polydisperse polysaccharide consisting of nonionic polymer molecules each made up of about 2000 monosaccharide units. Lower galactose substitution decreases the stiffness of the linear molecule and increases its flexibility. It also increases the extensibility of isolated molecular chains. The average molecular length of LBG is less than that for guar, which is ~7 nm. Overall, the grafted galactose residues in LBG molecules prevent strong molecular chain–chain interactions, but when there are up to 10 or more unsubstituted mannose residues in a row, time-durable junction zones might still be formed between such bare regions of molecular backbone. When the bare regions consist of only up to six mannose residues (such as those in guar gum), the junction zones are not durable and the molecular aggregate of polysaccharide can dissociate upon heating. This results in inducing water solubility. These nanocrystalline links dissociate more in hot water. Statistically, if the galactose residues were perfectly randomized or blocked, it is likely that each LBG molecule should have more than four such areas capable of acting as junction zones, which can allow gel formation.

As has been discussed in Chapter 2, galactomannan polysaccharides, occurring in a wide range of legume plant seeds, have some common structural features. These features consist of a linear, $\beta(1\rightarrow4)$-linked, linear, molecular backbone made up of D-mannopyranose units, to which short grafts of single, $\alpha(1\rightarrow6)$-linked D-galactopyranose are randomly attached. LBG galactomannan from the carob tree also has these structural features common with other legume galactomannans. The average M:G ratio in LBG has been generally described as 4:1, though it is closer to 3.5:1. This ratio in LBG can vary from 2.8:1 to 4.9:1 in the polysaccharide products from carob plants sources in different regions.

8.15   FINE STRUCTURE OF LBG[13]

In a galactomannan polysaccharide, the distribution of galactose grafts along the mannan backbone is referred to as the *fine structure* (Chapter 2). In general the galactose grafts on the mannan backbone are placed randomly, or in blocks, and not in a well-defined sequence, as was thought earlier. McCleary and coworkers have studied the distribution of galactose grafts on mannan backbone for guar and for carob polysaccharides. The study involved hydrolyzing of the polysaccharide using highly purified β-mannase enzymes from the microorganism *Aspergillus niger*.

It has been found that for the cleavage of the mannan backbone in a galactomannan by a β-mannase, which is an endoenzyme, there should be at least four unsubstituted mannose present continuously on the backbone. From the composition oligosaccharides produced in such enzymatic hydrolysis of purified LBG and by applying statistical methods, it was concluded that galactose grafts on the backbone are clustered into blocks, which can be as large as 25 or even more galactose residues being present continuously. Even larger portions of the mannan backbone are left unsubstituted. Blocks of continuously substituted backbone were called the *hairy regions*, while the unsubstituted regions were called *nonhairy, bare*, or *smooth regions*. From these studies, it was also concluded that LBG polysaccharide molecules also have a high proportion of galactose couplets but not many triplets as grafts. It was further concluded that blocks in which alternate mannosyl residue carried a galactose graft is very low. The proportion of blocks of unsubstituted D-mannosyl residues of the backbone appears to be similar to a galactomannan having random distribution D-galactose grafts.

Presence of large hairy blocks (25 or more continuous mannose) having galactose grafts, and even larger nonhairy blocks on the polymer chains has conferred certain unique functional properties to LBG, which are not shared by guar gum. Many other galactomannans, which have M·G ratios close to or even lower than LBG, do partially share such functional behavior with LBG. These unique structural features and functional behavior of LBG makes it indispensable for certain food applications, which we will discuss later.

8.16 MOLECULAR CONFORMATION OF LBG

LBG, like most of the other polysaccharides, is polydisperse. It consists of nonionic molecules, each containing ~2000 hexoses (D-mannose + D-galactose). Low galactose substitution of the linear mannan backbone results in increased flexibility of the polymer chain. LBG molecule is far less stiff when compared to the molecules of fenugreek or guar polysaccharide. Flexibility of LBG molecule, in turn, increases the extensibility of an isolated molecular chain.

It is known that more frequent grafts of galactose residues on the backbone of guar polysaccharide normally prevent strong chain–chain interactions. But when there are a large number of unsubstituted mannose residues in a row (as in case of the LBG molecule), junction zones are readily formed between different molecular chains. This results in partial crystallinity of LBG. When heating an aqueous solution of LBG, these nanocrystalline links (referred to as *junction zones*) can dissociate, resulting in increased solubility of LBG. Theoretically, it has been predicted that when the placement of galactose grafts on the molecular backbone is completely randomized into blocks, it is likely that each LBG molecule could have more than four nonhairy regions capable of forming junction zones and allowing gel formation to take place.

It has been mentioned that nongelling or weakly gelling polysaccharides (carrageenan, xanthan, and agar) have a partially organized, helical structure even in aqueous solutions.

At lower than room temperature (i.e., in the range of 0°C to 5°C), a number of the molecular polysaccharide chains can get reorganized further into a three-dimensional structure due the formation of junction zones. However, the gels thus produced are extremely weak and have a tendency to again melt at a room temperature of 25°C–35°C.

On addition of LBG at a low concentration to such gels, it can reinforce the already existing junction zones. This happens due to interaction of the helical structure of a weakly gelling polysaccharide and the linear, nonhairy regions of otherwise nongelling LBG. Thus, more firm binary gels can be produced from a mixture of two nongelling, or one nongelling and a weakly gelling polysaccharide. Relatively large and bare regions of the mannan backbone on the linear LBG chain can interact with several helices of the other polysaccharide to reinforce and stabilize a three-dimensional gel network. The liquid, imbibed into the unorganized portions of such a structure, can hydrate and immobilize the fluid into a gel.

8.17 SOLUBILITY BEHAVIOR OF COMMERCIAL LBG

The molecular weight estimation of undegraded LBG has been reported to be in a range of 300,000 to 360,000 dalton, corresponding to an average presence of nearly 2000 hexoses in a single molecule. Like most of the reserve plant polysaccharides,

TABLE 8.4
Solubility Behavior of LBG
Galactomannan Fractions
Having Different M:G Ratios

Fraction	Amount Weight (%)	M:G
Cold-water soluble	35	1.2:1
Hot-water soluble	52	3.3:1
Insoluble	10	5.2:1

LBG is also polydispersed, being composed of fractions having variable molecular weight and variable M:G ratios. Solubility of such a polysaccharide with bis-equatorially linked hexoses shall depend upon the number of grafts on the backbone and their distribution along the backbone.

If we consider an idealized structure of LBG having a M:G ratio 4:1, then it should be represented by a repeat unit of four β(1→4)-linked mannose, with every fourth mannose unit carrying a α(1→6)-linked D-galactopyranose group as a graft. This is shown next.

-(1→4) - β- D- Man- (1→4) - β- D- Man- (1→4) - β- D- Man- (1→4) - β- D- Man-

　　　↑

　(1→6)-α-D- Gal

Undegraded guar galactomannan has a molecular weight of nearly 1 million (~10^6 dalton). This is nearly 10 times the molecular weight of LBG. Guar galactomannan is differentiated from LBG as being cold-water soluble. For industrial applications, LBG is described as hot-water soluble and its solutions are always prepared in hot water (70°C–100°C). Reality concerning these polysaccharides is somewhat different. This has been shown by fractionation of both guar gum and LBG, which are described as cold-water soluble, hot-water soluble, and hot-water insoluble fractions. The M:G ratios found in these fractions are given in Table 8.4, which shows variation of this ratio can be from 1.2:1 to 5.2:1. Even in these fractions, the galactose stubs are not uniformly distributed.

8.18 PREPARATION OF LBG SOLUTION OR PASTE AND STRUCTURE–SOLUBILITY CORRELATION

Most β(1→4)-linked glycanes such as cellulose and mannans are known to have strong interchain hydrogen bonding interactions between the adjacent and parallel linear chains, which result in the formation of partial crystalline regions and poor water solubility. For galactomannans having a lower percentage of galactose and extended, nonsubstituted regions of the backbone, such as those in LBG, the unsubstituted portions of the backbone can come closer and "zip" together by hydrogen bonding. This can result in poor solubility, or solubility only in hot water, and even insolubility in some extreme cases.

If LBG is dispersed into cold water, at a concentration of ~5%, it remains as insoluble flocks, which settles down on standing. On the other hand, when LBG is added to hot water at a temperature of >70°C or when its dispersion in cold water is heated to above 70°C, the gum partially dissolves and some viscosity develops. Any insoluble portion of LBG is then prevented from separation and precipitation because of the viscosity developed due to the dissolved, soluble portion. Such pastes containing part dissolved and part suspended gum are generally suitable for use in the usual food applications of LBG.

Some researchers have suggested that when the polymer backbone has six or more continuous bare mannose on different chains, it may develop partial crystallinity due to hydrogen bonding. These hydrogen bonded regions open when the temperature of a solution is raised, which causes vibrational motion of the polymer chains. In the case of LBG, the crystalline regions formed due to hydrogen bonding between much larger unsubstituted portions, which can be as large as 25 bare mannose sugar units, are not disrupted by heat. These portions are poorly hydrated and part of the polysaccharide remains suspended in the viscous LBG paste produced by its partial dissolving.

For the purpose of viscosity measurements, LBG pastes are always prepared in water heated to ~90°C. The powder is then sprinkled into hot water, with vigorous stirring. Alternatively, the gum is dispersed in water at room temperature and then heated to ~ 90°C for 15–20 minutes and then cooled to 25°C before viscosity is measured.

8.19   SYNERGISM OF VISCOSITY AND BINARY GEL FORMATION

LBG by itself only forms very weak gels when its sol is cooled below 5°C. As such it does not have applications as a good gelling agent in food. It only forms good binary gels with certain nongelling or weakly gelling polysaccharides, or enhances the gel quality of other gelling but costly polysaccharides when used at a lower concentration. As an example, at a nongelling concentration of 0.5%, the seaweed polysaccharide carrageenan forms a good and firm gel on addition of 0.25% LBG to it. Addition of only 0.1% LBG to a carrageenan solution of 1.0% concentration results in an increase of the gel strength from 214 to 555 units. Similarly in the case of an xanthan–LBG blend, maximum gel strength is obtained, when the total polysaccharide concentration is 1.0 % and the ratio of xanthan to LBG in the blend is 7:3. Such blends of LBG with other polysaccharide gums have found applications in making low-sugar jams.

Unlike the blends with LBG, corresponding blends of carrageenan and guar gum does not form any gels. Only some synergistic viscosity increase is observed with guar gum.

8.20   BINARY GELLING SUBSTITUTES OF LOCUST BEAN GUM

Limitations in increasing the production of LBG have been discussed earlier. To meet the ever-increasing demand of LBG in the food industry, there is a need to find some equivalent hydrocolloid as its substitute. To some extent this can be met by discovering other galactomannan gums having a lower percentage of galactose.

Chemotaxonomy studies of Leguminoseae plants based on their M:G ratio have shown that, in general, the galactomannans from full-grown legume trees or shrubs (carob and tara) have a lower percentage of galactose compared with annual crops (guar and fenugreek). The M:G ratio in galactomannans determines the functional properties of any gum. Since LBG has the optimum M:G ratio and distribution of galactose grafts on the mannan backbone, the use of LBG in food is not likely to diminish except for price consideration and that too with some loss in the quality of food products.

Lab scale attempts toward reducing the M:G ratio in guar gum using a purified α-galactosidase enzyme to selectively cleave some galactose grafts and bring it down to the same level or even to a lower level than LBG did not result in required binary gelling property in guar. Hence, LBG continues to be used in specific food applications, such as in ketchup, ice creams, soft cheese, yogurt, and other dairy products. Thus, LBG cannot be replaced completely by guar gum as an additive in food. An ice cream stabilizer based on a blend of guar gum and LBG has a better meltdown behavior than that having either guar or LBG alone. In some applications the use of tara gum has been made to replace guar–LBG blends.

A gum with a higher percentage of galactose has good cold-water dispersibility and higher viscosity but poor gelling property. This observation can form a guideline for exploitation of new galactomannans of the desired M:G ratio. M:G ratios for some edible galactomannan gums is as follows: LBG or carob gum, 4:1; tara gum, 3:1; guar gum and Sasbania bispinosa gum, 2:1; and fenugreek gum, 1:1.

8.21   SUBSTITUTES FOR LOCUST BEAN GUM

The oldest use of LBG has been as an additive to pulp for making paper. Various additives, including LBG, were mixed with wood pulp for making paper, which resulted in the enhancement of bonding and coherence between fibers. To increase the ultimate dry strength of paper, materials most commonly now used are starches, polyacrylamide resins, and guar gum. The most commonly used starches being used is the modified type known as cationic starch. When dispersed in water, cationic starch assumes a positive surface charge. Because the cellulose fibers in the pulp normally have a negative surface charge, there is an affinity between the cationic starch and the cellulose fibers. The natural cellulose interfiber bonding that develops into a sheet of paper as it dries out is considered to be due to inter- and intramolecular forces of attraction, which are called the *hydrogen bonding*. Since the cationic starches have proven to be cheaper and better additives for paper, the use of LBG in paper has now been completely abandoned.

8.22   CROSS-LINKING REACTIONS AND INTERACTIONS
OF LBG WITH OTHER POLYSACCHARIDES

The mechanism of cross-linking reactions and the applications thereof for galactomannans were discussed in Chapter 7 (guar gum). In common with other galactomannans, and by virtue of the presence of *cis*-hydroxyl groups in the constituent sugars, LBG can also undergo strong hydrogen bonding interactions and cross-linking reactions. Thus, LBG can be gelled, reversibly by borax at an alkaline pH and also form complexes with transition metal ions (e.g., zirconium and titanium).

Fine particles of LBG have been surface cross-linked by dialdehyde (e.g., glyoxal) to make it nonlumping. Such cross-linking involves bis-acetal formation by the *cis*-pair of hydroxyl groups of unsubstituted mannose backbone of two or more adjacent molecular chains. Acid catalyzed, glyoxal bis-acetal formation takes place in the presence of an acid (at pH ~3–3.5) and it can render fine-mesh LBG powder dispersible without lumping. Full viscosity in such a case only develops when the pH of the dispersion is adjusted to ~7, that is, when the bis-acetal cross-linking is reversed. Extensive industrial applications of such reversibly cross-linked galactomannans are made in industries where very large volumes of galactomannan pastes are used, for example, in oil-well drilling.

Like most of the plant hydrocolloids, aqueous solutions of LBG at a concentration of 0.2%–2.0% are shear thinning and pseudoplastic. LBG sol can show thixotropic and semigel rheology at higher concentrations. Viscosity of an LBG solution decreases with an increase in concentration of alcohol in an aqueous solution of LBG. At a sugar concentration of ~60%, LBG forms a weak gel. Under similar conditions, a precipitate is formed in the case of guar polysaccharide.

The initial viscosity of 1%, high-grade LBG paste, which is made in hot water at 95°C is lower than that when it is cooled to 25°C for the final viscosity measurements. The viscosity comes to 400–800 in cps units depending on the purity grade of LBG. The measurements are made using a Brookfield synchro-lecteic viscometer. This is the normal behavior for most galactomannan polysaccharides in that the viscosity decreases with a rise in temperature. For a particular LBG sample, the viscosity decreases with an increase in shear rate (i.e., the revolutions per minute at which the viscosity is measured).

At room temperature (~30°C), only 35% of the cold-water soluble LBG fraction dissolves in water to develop a very low viscosity. When heating this dispersion, the viscosity due to the dissolved cold-water soluble fraction decreases. At a temperature higher than 45°C, the hot-water soluble fraction of LBG starts dissolving with a further increase in viscosity, which attains a maximum value of ~450 cps at 80°C. This viscosity is far less when compared to the much higher viscosity (~5000) of guar gum at the same concentration. When the temperature is lowered to ~5°C, a weak gel is produced. Hence, invariably LBG pastes are prepared at a temperature high than 80°C and then the sol is cooled to the temperature at which the viscosity measurement is made.

At a concentration of 0.5%, the viscosity of the LBG solution, which is prepared as usual at 80°C–90°C, followed by cooling to room temperature should be extremely low (i.e., around 50 cps), whereas a high grade of LBG at 2.0% concentration should have a viscosity of 4000–8000 cps. Some loss in viscosity can occur on shearing at a high rate for longer periods, which is due to molecular breakdown.

There is a strong interaction between LBG and other polysaccharides, typically carrageenan, agar, and xanthan. Thus, these polysaccharides form more firm binary gels in the presence of LBG or can gel at otherwise nongelling concentrations.

Formation of these binary gels has found many industrial applications of LBG for making gelled food, which has proved to be economical.

The following equation has been suggested as a correlation between molecular weight and intrinsic viscosity of the LBG solution:[11]

$$[\eta] = 9.28 \times 10^{-6}. M^{0.98}$$

8.23 MODIFICATIONS AND DERIVATIZATION OF LBG

Whereas chemical derivatization of guar gum is being practiced on a large commercial scale, LBG derivatization has not been carried out on an industrial scale because of the initial high cost of LBG. Further, in contrast with guar gum, not much is gained as regards to the functional properties of LBG by derivatization. LBG can be rendered cold-water soluble by derivatization, such as carboxymethylation, phosphorylation, and hydroxalkylation. General methods for derivatization of galactomannans and other polysaccharides have already been discussed elsewhere (Chapter 3), and these are applicable to LBG also. Initial heating of LBG after blending with sugar or guar gum forms cold-water soluble LBG. In this process the crystalline regions of LBG are opened and these are not reformed rendering LBG cold-water soluble.

8.24 APPLICATIONS AND USES OF LBG[5]

In most of the literature on guar gum and LBG, one can find that the listing of the applications of these two gums is almost identical. In the patented literature, several applications of LBG for nonfood industries have been described. After its commercialization in the late 1950s, guar gum proved to be more economical compared to LBG in most of the nonfood applications.

Currently, in most of the nonfood applications of galactomannans only guar gum is being used. This is due to the fact that guar gum is much cheaper and more abundant compared with LBG. Yet mentioning of such technical applications of LBG has become an academic routine.

Nonfood and industrial applications of LBG. Long before the emergence of guar gum as an industrial polysaccharide, there have been many nonfood industries using LBG. Thus, LBG and its derivatives were extensively used in the U.S. paper industry as a wet end additive. In the textile printing industry, the soluble hydroxyethyl and anionic carboxymethyl LBG were used as printing paste thickener. LBG derivatives were used in oil-field applications as a thickener for drilling mud as well as for oil-well fracturing to enhance oil and gas recovery. In most of these oil-field applications, guar gum has completely replaced LBG. This is also true of many other nonfood applications of LBG, for example, its use as a textile printing paste thickener, as a moisturizer for tobacco products, and for manufacturing of slurry explosives. Guar gum has almost completely replaced LBG in most of these applications.

Wherever the functional properties of guar gum permitted replacement of LBG, it was always done because of the price considerations and because of the limited production of LBG. Higher price and limited availability has severely restricted the applications of LBG in nonfood applications. Guar galactomannan polysaccharide can now be used effectively and economically in those applications where LBG was earlier used.

Clearence of LBG as a safe food additive and its price structure. Since the late 1950s, in the United States and most European countries, the major consumption of LBG has only been in the food industry, and its annual consumption has been relatively stable at ~12000 tons. This is about 75% of total LBG produced. In contrast, only about 40% of annual world production of guar gum is used in the food industry, while the rest of it has technical applications. In the case of guar gum and LBG, the U.S. Food and Drug Administration and similar bodies in most of the other countries have cleared these two polysaccharides as a GRAS class food additive (Title 21 Code of Federal regulations, part 184.1343).

The 2007 market price of LBG was $10–$15/kg, which was approximately $6–$7/kg in 1985. In comparison to this the price of food-grade guar gum has held steady at $1.50–2.50/Kg). Even the use of guar gum or any other cheaper hydrocolloid rather than LBG is always preferred wherever possible in the food industry. This has seriously limited the use of LBG in even food applications. LBG is now used only where the use of guar gum (in food) has certain limitations, for example, not having a semi-gel rheology.

The discussion on the difference in the rheology of LBG and guar gum clearly indicates that in some applications at least (e.g., for binary gelling and frozen foods, cheese products, yogurt, and sour cream) guar gum certainly cannot completely replace LBG. Unique swelling, heat-shock resistance, and water-binding properties of LBG make it indispensable in diary and other food applications. Guar, which is only a viscosifier, certainly cannot replace LBG or tara gum in applications where gelling is required, and in such food applications LBG is preferred.

Interactions of LBG with other polysaccharides and binary gelling. It has already been mentioned that binary gelling polysaccharides (e.g., carrageenan, xanthan, and agar) have a partially organized, or a tertiary, helical structure in solution, even under nongelling conditions. On lowering the temperature, the molecular chains in these organize further into a gel due to the formation of a three-dimensional structure and creation of junction zones between the molecular chains. Addition of LBG at low concentration reinforces these junction zones by interaction of the helical structure of gelling polysaccharide and the long bare regions of the mannan backbone. In such interactions, a single LBG chain can interact with several helices of other polysaccharides to reinforce and stabilize a three-dimensional gel network having liquid imbibed into the unorganized portions of the structure. The thickened fluid is thus converted into a gel.

As a typical example, at a nongelling concentration of 0.5%, carrageenan forms a firm gel when 0.25% of LBG is added to it. Similarly, in the presence of 0.1% LBG and at 1% concentration of carrageenan, the strength of gel produced increases from 214 to 555 units. Similarly in case of xanthan blended with LBG, maximum gel strength is obtained when the total polysaccharide concentration is 1% and the ratio of xanthan to LBG in the blend is 7:3. Thus, the use of LBG blended with several costlier polysaccharides has proved economical in gelled food production.

Dairy and frozen desserts. A major portion of LBG being used as a food additive is in dairy products, where it is used, along with guar gum, xanthan, and carrageenan, as a stabilizer in ice creams and flavored milk products. LBG, when used at concentrations of 0.05%–0.25%, alone or along with other gums, provides excellent heat-shock resistance, improved meltdown behavior, superior texture, chewiness, and emulsification in many food products. It is used in soft cheese, cheese spread, curd, and yogurt, where it prevents water separation and produces resilient body and texture. When used, along with carrageenan, LBG is very effective in preventing separation of casein proteins in milk products, even at acidic pH, for example, in the flavored and spiced milk products.

Meat products and pet foods. LBG has been used as a binder in meat products and pet foods where, acting as a lubricant, it makes extrusion and stuffing easier. It improves product homogeneity, moisture retention and texture, and increases the shelf life of most food products. It is an excellent stabilizer for a variety of hot-packed canned products. LBG is used in meat and canned pet foods, extruded food products, bakery products, soft confectioneries, and dietary foods.

Bakery products. LBG has been extensively used in the bakery industry. Thus, it has been used as a food additive in the flour used in the manufacturing of bread, cakes, biscuits, and other baked goods. It acts as a moisture retention aid, increases duff volume, improves the texture and shelf life of a baked product, and reduces crumb formation. Being more economical, food-grade carboxymethylcellulose (CMC) and guar gum are two strong competitors of LBG in these applications.

Miscellaneous applications. LBG has been used in making low calorie, dietary products for diabetic and obese persons, where it acts as a soluble dietary fiber for modulating sugar and lipid metabolism. It is used in confectionery products, along with xanthan gum, to produce gelled products. Such gelled products are also used in room air-fresheners. It is used in certain medicinal, pharmaceutical, and cosmetic products, for example, toothpastes, lotions, and creams.

REFERENCES

1. Duke, A., ed., Handbook of Legume of World Economic Importance, Plenum Press, New York, 1981.
2. Loock, E. E. M., The carob or locust tree, *Ceratonia siliqua*, J. S. Afr. For. Assoc., 4 (1940): 78–80.
3. Whistler, R.L. and BeMiller, J. N., eds., Industrial Gums, 3rd ed., Academic Press, New York, 1993. (2nd ed., 1973)
4. Coit, J. E., Carob culture in the semi-arid southwest, Rittenhouse, San Diego, CA, 1949.
5. Davidson, R. L., ed., Handbook of Water-Soluble Gums and Resins, McGraw-Hill, New York, 1980.
6. Reid, J. S. G., Galactomannans from legumes seeds endosperm, Adv. Bot. Res., 11 (1985): 125–155.
7. Anderson, E., Ind. Eng. Chem., 42 (1949): 288.
8. Buffington, L. A., Stevens, E. S., Morris, E. R., and Rees, D. A., Int. J. Biol. Macromol., 2 (1980): 199.
9. Dea, I. C. M., Morris, E. R., Rees, D. A., Eelsh, E. J., Barnees, H. A., and Price, J., Carbohyd. Res., 57 (1977): 249.

10. Dea, I. C. M., Interaction of ordered polysaccharides. Structure-synergism and freeze-thaw phenomenon, Proc. Easter School Agr. Sci., University of Nottingham, (1979): 229–247.

11. Dea, I. C. M., In Gums and Stabilizers for Food Industry, G. O. Phillips, P. A. Williams, and D. J. Wedlock, eds., IRL Press, Oxford University Press, Oxford, 1990, 373.

12. Pederson, J. K., Cereal Sci. Today, 10 (1974): 476.

13. Dea, I. C. M., Morrison, A., Adv. Carbohyd. Chem. Biochem., 31 (1976): 241.

9 Fenugreek Gum
The New Food Galactomannan

9.1 INTRODUCTION

Among agricultural crops that produce grains for use as human food, legumes or the pulses stand next only to the grasses, or Gramieae family, which include wheat, barley, rye, rice, and maize.[1] A major constituent of human food is various types of carbohydrates, present in edible seed grains. The main polysaccharides from the edible seed grains of the Gramieae family are starches and cellulose. Various sugars, particularly sucrose, are present in the stem of sugar cane (Gramieae family) and sugar beet, which is a tuber. Legumes produce more varied types of polysaccharides and many more oligosaccharides than those produced by the seeds of grasses.[2,3] Thus the sugars, which the grasses mainly produce, are sucrose, glucose, and fructose, whereas the polysaccharides produce cellulose and starches. Starch is the major reserve polysaccharide in grasses and some tubers (e.g., potatoes), whereas cellulose is their structural polysaccharide. Cellulose is insoluble and nondigestible in the human system. Besides these some hemicelluloses are always present as a binder for cellulose fibers.

Many legume seeds have galactomannans as the endosperm reserve polysaccharide. Legume seed used in human food are typically cooked in India, as a *dal* (an Indian recipe), or several other recipes based on them. Each of these recipes has a different and characteristic taste and flavor. This is due to the occurrence of several minor polysaccharides in the legume seeds. These polysaccharides impart specific taste and flavor to each of the legumes.[3] Being variable and minor constituents in different legumes, and not the major reserve polysaccharide unlike galactomannans, many of these polysaccharides have not been isolated or studied in detail for their chemical structures.

Guar and fenugreek[4,5] are the two major galactomannan-producing annual legume crops. There are also many legume trees—such as carob, tara, and *Cassia fistula*—that produce seeds in their pods that contain galactomannans as the reserve polysaccharides. Because of their unique applications in food as well as in nonfood industries, legume seed galactomannans have acquired great commercial value. There are still many underutilized, tropical legume galactomannans that could be exploited as future human food resources in developing countries.

A panel of scientists and advisory committees on technology innovation of the National Academy of Sciences in the United States, in collaboration with the United Nations Organization (UNO), has made a compilation of underexploited legumes as a resource for the future[5] and have published a database on legumes of eight South Asian countries. The National Botanical Research Institute at Lucknow (CSIR, India) has also published similar data on legumes of the Indian subcontinent. These databases are helpful in getting information about many lesser exploited yet promising legumes as a source of polysaccharides and other agroproducts.[6]

India has been fortunate to have many underexploited, natural plant resources and it is time that the country starts undertaking developments on utilization of such resources. Fenugreek is one such annual legume crop, whose strongly scented seeds are a source of unique galactomannan in addition to other useful constituents.[2,7,8]

9.2 FACTORS THAT INFLUENCE COMMERCIALIZATION OF POTENTIAL POLYSACCHARIDE GUMS[7]

According to Whistler and BeMiller,[7] any newly investigated plant polysaccharide gum, or hydrocolloid, is likely to be adopted for industrial production and use when

1. It has some specific functional properties, such as being a good viscosity builder, or a good gelling agent, for which it is being considered for commercialization.
2. Gum cost is low and it is likely to remain reasonably constant for several years to come. As regards to the cost factor, the galactomannans from annual crops (e.g., guar and fenugreek) are cheaper compared to those from full-grown trees (e.g., carob), and the supply of gums from annual crops is more sustainable and assured.
3. The supply of a high quality of a product (gum) from a particular plant source is well assured, particularly when there is an increased world demand for it. This again is true of gums from land-cultivated annual crops rather than those from perennial trees, seaweeds, and microbial sources, which also produce some unique polysaccharides.
4. When a gum is meant for human consumption, consideration for its acceptance as a food, cosmetic, or drug additive by the appropriate government agencies, such as the U.S. Food and Drug Administration (FDA), becomes an important consideration. In many cases, a physically modified, partially depolymerized, or chemically derivatized gum provides better functionality compared to the natural gum from which it is derived.

Several water-soluble or dispersible, high molecular weight galactomannan polysaccharides function as hydrocolloids and fenugreek polysaccharide is one of them.

About 20 years back, only the following three, commercially produced legume seed endosperm galactomannans were approved as safe food additives:

1. Guar gum (from *Cymoposis teteragonnolobous*, or the guar plant)
2. Locust bean gum (from *Ceratonia siliqua,* or the Carob tree)
3. Tara gum (from *Caesalpinia spinosa,* or the tara tree, also known as *huarango* or *guararnga* or the Peruvian carob tree)

Commercially isolated, fenugreek seed galactomannan has been a new addition to this list of edible galactomannan gums. Since the whole seed of fenugreek is traditionally considered to be edible, its endosperm mucilage did not need fresh clearance as an edible gum from government agencies.[8] Of the aforementioned galactomannan-bearing plants, guar and fenugreek are annual crops mainly cultivated on the Indian subcontinent. Locust bean gum and tara seed gums, which are described in Chapters 8 and 10, respectively, come from evergreen perennial trees or shrubs, which are grown in the costal Mediterranean sea regions of Europe and Africa, and Peruvian Andes in South America, respectively.

9.3   FENUGREEK GALACTOMANNAN AS AN EMERGING INDUSTRIAL POLYSACCHARIDE[9,10,11]

The name *fenugreek* comes from Latin meaning "Greek hay." In different regional languages of India, where fenugreek is extensively grown and used as a spice food additive, it is called *mentya* in Kanada, *vendayam* in Tamil, *menthulu* in Telgu, and *methi* in Hindi and in Urdu.

Fenugreek (*Trigonella foenum-graecum*) is an annual legume plant native to the Mediterranean region. From ancient days, it has been grown in the Near and Middle East regions, Africa, India, and several other parts of the world, where it is used as a spice food additive and for some medicinal applications. It is now grown all over the world, including Europe, Canada, and the United States.

According to Whistler, who is an authority on gums,[7] fenugreek seed endosperm galactomannan was not in industrial production until 1993, when he suggested

> there exists a strong incentive to make dual use of the fenugreek seed by removing the culinary spice and other components and separate its unique galactomannan, as well as its herbal constituents, which contains steroidal saponin, "disogenin." Disogenin is in good demand for making the sex hormone (cortisone), which is used for making oral contraceptives.

Like several other prophecies regarding the future prospects of many underutilized polysaccharide sources made by Whistler, this prediction was soon to be fulfilled. Commercial production of fenugreek gum is currently (year 2000 onward) being done by some industries in Europe, Israel, India, and the United States. Some trade names of fenugreek gum being marketed are FenuLife, FenuPure, and Air Green.

Commercial production of a galactomannan polysaccharide from fenugreek seed, which has been undertaken nearly 40 years after guar gum, which was first manufactured in the mid-1950s. Some industries have now started producing and marketing a sizable amount (an exact figure is not yet available) of fenugreek gum. Other fenugreek products are also finding increasing applications.

Since fenugreek is a widely grown annual agriculture crop, a sustainable supply of its seed is well assured. As mentioned earlier, besides the gum, fenugreek seed also contains spicy oil, saponins, and a good quality of lysine-rich edible protein. The presence of all these constituents make fenugreek a cost-effective agricultural crop.

Considering all these factors, the Research Branch of Agriculture Canada[9] has started a project to produce a fenugreek crop of an improved seed yield. A breeding program to produce seeds yielding higher levels of saponins from fenugreek seed was also undertaken in England, but no specific breeding program to improve gum yield from fenugreek seed appears to have been undertaken so far. Improvement of seed quality and its yield can naturally result in an increased yield of gum and most of the other seed components and value of the crop.

9.4 FENUGREEK CROP[11]

The fenugreek plant is a slender annual herb of the pea family (Fabaceae). Its dried seeds are used as a spice food, flavoring, and medicine. The seeds have a strong aroma and taste. The seeds have a mixed sweetish and somewhat bitter, reminiscent taste of burnt sugar and have a farinaceous texture. These may be mixed with flour for bread or eaten raw or cooked. The herb is a characteristic ingredient in some curries and chutneys, and is used to make imitation maple syrup. In India, young fenugreek plants are used as a potherb. In Northern Africa the plants are used as fodder. Traditionally considered an aid to digestion, the seeds have been used as an internal emollient for inflammation of the digestive tract and as an external poultice for boils and abscesses.

Native to Southern Europe and the Mediterranean region, fenugreek is cultivated in Central and Southeastern Europe, Western Asia, India, and Northern Africa. Thus, it is now cultivated worldwide as a semiarid crop. The plant grows in well-drained soil under a mild climate. The seed crop matures in 3–4 months and yields 300–400 Kg seeds per acre. The cultivation of the fenugreek crop in the American continent has been increasing due to its high content of water-soluble galactomannan gum, which is a dietary fiber. Fenugreek cultivation is likely to increase further due to the current increasing demand of its gum, which has been found to regulate blood glucose and insulin response in normal and diabetic persons. Another unique property of this galactomannan is its action in stabilizing oil–water emulsion.

As an annually cultivated spice bean crop in France, Egypt, and Argentina now extensively grown in India, which is a major fenugreek seed exporting country in the world, follow it. As the cultivation of guar crop increased due to commercialization of guar gum, similarly the cultivation of fenugreek has also increased dramatically during the past two decades due to the commercialization of its gum.

The fenugreek plant is erect, loosely branched, and about half a meter tall with trifoliate, light green leaves, and small white flowers. The slender and sword-shaped pods are up to 15 cm long, curved and beaked, and contain 10–20 yellow-brown seeds. The oblong, 2–3 mm long seeds are hard, with an outer, wrinkled seed coat. In India, fresh and green, slightly bitter and spicy, fenugreek leaves are used as a green leafy vegetable. The strongly scented fenugreek seeds are used as a spice to flavor curries (e.g., *sambahar*), pickles, and chutneys in India and other neighboring countries. In Canada it has been used for making imitation maple syrup.

The seeds of the fenugreek herb are also an effective nutritional supplement and have been used by herbalists for many centuries for the health benefits it provides. According to the ayurvedic and unani systems of medicine, fenugreek seed has been used in some condiments and as a spice component of food in India, and Middle and Far East for centuries. Because of its being nontoxic and its innocuous nature, fenugreek seed has traditionally been used as a constituent of food, and no FDA approval for the use of its isolated gum was ever required.

Thus, fenugreek is an amazing, magic herb that can cure a number of ailments. Indian Ayurvedic and traditional Chinese medicine recommend fenugreek to treat arthritis, asthma, bronchitis, improve digestion, maintain a healthy metabolism, increase libido and male potency, cure skin problems (wounds, rashes and boils), treat sore throat, and cure acid reflux. Fenugreek also has a long history of use for the treatment of reproductive disorders, to induce labor, to treat hormonal disorders, to help with breast enlargement, and to reduce menstrual pain. Recent studies have shown that fenugreek seed powder helps to lower blood glucose and cholesterol levels, and may be an effective treatment for both type 1 and 2 diabetes. This is due to its galactomannan gum. Fenugreek is also being studied for its cardiovascular benefits.

9.5   COMPOSITION OF FENUGREEK SEED[8,12]

Dicotyledonous fenugreek seed has a wrinkled brown-yellow seed coat or the husk, enclosing two yellowish-white, translucent endosperm halves, which are mainly composed of a water-soluble galactomannan polysaccharide or the gum. Between the endosperm halves of fenugreek seed, a yellow germ portion is sandwiched. Fenugreek seeds are rich in several nutritional components and have been used by herbalists for many centuries for the health benefits it provides. The germ portion is mainly composed of a good quality of edible protein and ether extractable oil (7%–9%), which is a mixture of fatty oil and flavoring essential oil. The alcohol extractable (5%–7%) portion of the seed consists of saponin (disogenin), a yellow flavonoid pigment, alkaloid (trigonelline), some free amino acids, L-lysine and L-tryptophan, and minerals (potassium). It also contains vitamins of the B complex group and vitamin C.

Among bioactive constituents present in fenugreek seed, yellow flavonoid pigment has strong antioxidant properties. Diosgenin in fenugreek is thought to help increase libido and lessen the effects of hot flashes and hormone-induced mood fluctuations. It contains phytoestrogens, which mimic the effects of the hormone estrogen. Seed also contains steroidal saponins (diosgenin, yamogenin, tigogenin, and neotigogenin) and mucilaginous fiber, which are believed to be responsible for many of the beneficial effects fenugreek exhibits. The mucilaginous fenugreek fiber (galactomannan gum) is believed to be responsible for many of the beneficial, health effects that fenugreek exhibit. Fenugreek seed also contains a nonprotein amino acid, 4-hydroxyisoleucine, which is reported to have the biological activity of reducing blood sugar. Insoluble fibrous material in fenugreek is mainly from the husk and consists of cellulosic fiber and pigments, and the endospermic galactomannan is a water-soluble gum.

The average composition of fenugreek seed is given in Table 9.1.

In germinating fenugreek seed, the endosperm galactomannan, which is the reserve polysaccharide, is used up as a carbon source during the growth of the plant embryo in the soil. Only when the plant emerges out of the soil, it starts photosynthesis using atmospheric carbon dioxide. Germinating fenugreek seeds produce

TABLE 9.1
Major Constituents of Fenugreek Seed

Seed Component	Amount (%)
Moisture	2.5–3.5
Protein	25–30
Lipids (ether extractable)	7–9
Steroidal saponins	5–7
Galactomannan	25–30
Insoluble fiber (cellulosics)	20–25
Ash	3–4

Sources: Garti, N., Madar, Z., Aserin, A., and Sternheim, B., Food Sci. Tech., 30 (1997): 305–308. Garti, N., Food Hydrocolloids, 8 (1994): 155; Mathur, N. K. and Mathur, G. M., Fenugreek gum, Science-Tech Entrepreneur, December 2006.

β-mannase and α-galactase enzymes, which can degrade (hydrolyze) galactomannan polysaccharides.

Just like other galactomannans, fenugreek gum has a strong tendency to bind and hold moisture to overcome water stress during its scarcity periods. When fenugreek seed dries up, there is considerable shrinkage in its size due to the loss of water and it results in the formation of wrinkles on the seed coat. Solvent defatted, deodorized, and protein-free fenugreek gum is a tasteless, odorless, slightly yellowish powder. Just like other galactomannans, it has been reported to be effective in lowering blood sugar and blood lipid level. Additionally it promotes growth of probiotic intestinal flora and acts as a soluble dietary fiber.

9.6 MANUFACTURING OF FENUGREEK GALACTOMANNAN GUM[11,12]

Important commercial galactomannan gums (e.g., guar gum and locust bean gum) are produced by stepwise dry grinding, sifting, and sieving of the seed components. The process results in effective separation of nearly pure and intact endosperm from the easily pulverized husk and germ components of the seed. This is followed by further grinding of separated endosperm into a powder of desired mesh size. Such dry processing gives high product yield and has generally kept the processing cost low.

Among the galactomannans, fenugreek gum is unique in having mannose and galactose in a nearly 1:1 ratio, and it is more cold-water soluble than any other galactomannan, including guar gum. In the case of fenugreek, the patented industrial production methods have described only the water extraction method of gum manufacturing. This may be due to the higher aqueous solubility of fenugreek

---

TABLE 9.2
Specifications of Fenugreek Gum
Currently Being Marketed

Constituents	Amount (%)
Acid-insoluble matter	3–5
Water-insoluble matter	2–3
Loss on drying (moisture)	8–10
Protein	3–5
Starch	Absent
Soluble galactomannan (dietary fiber)	70–85
Ash	1–2
Ether extractable material	1–2
Viscosity (1% aqueous solution)	500–2000 cps
Particle size	70–100 mesh
Calorie (energy value)	1 kcal/g

---

polysaccharide. Such wet-extraction methods are generally adopted from the general laboratory method of extraction of a water-soluble seed polysaccharide.

The wet-extraction methods, described in different patents,[13] differ only slightly in their details. The methods generally consist in course powdering of the whole fenugreek seed, followed by petroleum ether and ethanol extractions of the powder to remove fatty and flavoring oils, pigments, saponins, bitter-tasting components, and other low molecular weight substances. This is followed by hot-water extraction to dissolve the soluble galactomannan polysaccharides, leaving behind insoluble cellulosics from the husk and most of the germ proteins. These insoluble components are removed by filtration or centrifugation. From the clarified filtrate, the galactomannan polysaccharide is precipitated by adding alcohol up to 60% concentration, which is then filtered and dried. Redissolving of crude polysaccharide in water followed by a second alcohol precipitation yielded purer gum product. Alcohol used in the process was recovered by fractional distillation and reused.

For economic consideration, other marketable products of fenugreek seed are also recovered as mentioned in these patents. Typical specifications of fenugreek gum being marketed are given in Table 9.2.

9.7 DRY VERSUS WET METHODS OF EXTRACTION OF SEED GALACTOMANNANS[10,11]

A major portion of polymeric materials in any dry plant material consists of proteins, lignins, cellulosics, and soluble and insoluble polysaccharides. The cost of raw material and processing extraction of a commercially pure polysaccharide from any plant material can vary considerably depending on the process being used. Thus, the manufacture of many seaweed and microbial gums and pectin polysaccharides involve their extraction into water followed by filtration to remove insoluble portion. This is followed by precipitation of the water-soluble polysaccharide with alcohol, which is later recovered by fractional distillation for reuse and final drying of the precipitated polysaccharide. Among all the galactomannans, fenugreek gum has the highest water solubility, and this could be the reason for adopting a wet extraction method for its manufacture.

In contrast to such wet-extraction methods, most legume endosperm polysaccharides of commercial purity, which are galactomannans by nature, are generally separated by differential grinding, sifting, and sieving to separate the three main seed components. These are the seed coat, endosperm splits, and germ. Numerous patents for making various legume seed galactomannans have been granted. Most of these are purely mechanical operations, which keep the processing cost low.

According to the author, both dry and wet extraction of galactomannan seemed feasible for fenugreek. The author has devised a dry process (patent pending) for producing fenugreek seed gum, which is based on a slight modification of the currently used guar gum extraction method (Chapter 7) being used in India.

The method essentially consists in course grinding of seed in which moisture content should be 6%–8%. In case of well-dried and marketable fenugreek seed the moisture content is low (2%–3%), which is increased to ~7% by soaking the seeds in water and redrying. This is followed by splitting the seed into two endosperm halves by course milling. Dehusking of endosperm is carried out, as in the case of guar seed, and after sieving off the pulverized husk and germ, the purified endosperm is ground into a powder of desired fineness.

Fenugreek endosperm is soft compared to most other galactomannans bearing legume seed endosperm. This clearly demonstrates less molecular chain–chain interaction (hydrogen bonding) between completely galactose-substituted mannan backbones of this molecule. When the mannan chains are almost completely substituted by single-chain galactose stubs, their natural packing in the endosperm cannot be very tight. This explains softer fenugreek endosperm.

Dry processing of fenugreek seed yields gum of much higher viscosity compared to the wet process. This difference might be due to the fact that any hydrolytic breakdown of the polymer molecules taking place in the hot-water extraction process is largely reduced in the dry process.

For methods based on hot-water extraction of whole seed powder, the galactomannan hydrolyzing enzymes, which are present in the seed germ, can bring about some hydrolysis of the endosperm polysaccharides. This can result in reduced yield and gum powder of lower viscosity.

9.8 NEED TO ADOPT DRY PROCESSING FOR COMMERCIAL MANUFACTURING OF FENUGREEK GUM[11]

The author has felt a need to develop and optimize a suitable dry process of manufacturing fenugreek galactomannan from its seed, as has been done in the case of other legume seed polysaccharides. Whereas the gum recovered from fenugreek seed by wet processing is around 20% or even less, a suitable dry process can give a better yield of ~25%, as in the case of other legume seeds. Morphological structure of dicotyledonous fenugreek seed suggests that by mild-heat treatment, it should be possible to pop the husk from fenugreek splits. Such dehusking has been done in case of other legume seeds bearing galactomannans. This can be followed by course grinding, which the relatively hard endosperm generally can stand, while the more fragile germ gets powdered, and it is easily separated from the intact endosperm split by sifting.

Not much attention has been paid to develop such dry processing of fenugreek. It may appear that due to small seed size, separation of industrially pure endosperm from the husk and germ in fenugreek seed might not be effectively achieved. However, in the case of *Sesbenia bispinosa*, which is another seed that is smaller than guar seed, successful separation of seed endosperm has been achieved by employing machines slightly modified from those being used in making guar split.[3]

Fenugreek gum extraction by a dry grinding process, could create a problem due to the characteristic smell of its oil and the bitter constituents getting into the endosperm gum making it unpalatable. I have not encountered any such problem in dry processing of fenugreek gum. However, solvent washing of split to remove any odorous substances before grinding can solve the problem.

Highly pure endosperm split obtained by a differential grinding and dry process, which is based on a slight modification of the currently used machines being used to produce guar gum split, has been shown to produce an almost odorless product and without any bitter taste. The yield of gum is ~25%, which is better compared to that in a wet process.

9.9 CHEMICAL STRUCTURE OF FENUGREEK GALACTOMANNAN[13,14]

All the legume seed galactomannans consists of a β(1→4)-linked linear mannan backbone, to which single galactose grafts are linked randomly or in blocks by an α(1→6) glycoside bond (Chapter 2). Fenugreek gum also has this common structural feature of galactomannans.[15] Galactomannans from different legume seeds mainly differ in their mannose-to-galactose ratio (M:G), molecular weight, and mode of placement of the galactose grafts, which are generally not spaced regularly but placed randomly on the linear backbone. In case of fenugreek gum, no such investigation is needed, because nearly all the mannose units of polymer backbone carry a galactose stub.

Fenugreek polysaccharide has been reported to contain a small amount of sugars other than mannose and galactose.[16,17] This discrepancy might be due to the lack of good purification of the final product, resulting in contamination of the lab sample or due to the analytical method employed.

As reported, the M:G ratio in fenugreek gum samples, in most cases, is close to 1:1, which makes fenugreek gum as one of the highest galactose-containing galactomannan (~48%; and M:G = 1.02:1).[18] According to Madar and Shomer fenugreek galactomannan consists of two fractions, with M:G ratios of 1.5:1 and 1.1:1. Other than fenugreek, lucerne (*Medicago sativa*) and clover (*Trifolium pratense*) are two other, less common galactomannans, which have ~48% galactose.[19]

The linear mannan backbone of fenugreek gum has the usual α (1→6)-linked, single galactose grafts on nearly all the mannose units of the main chain.

Molecular weight of the polysaccharide has been estimated to be ~30,000 dalton. This corresponds to an average presence of 180–190 monosaccharides (mannose + galactose) units in a molecule. On an average, the linear mannan backbone of fenugreek polysaccharide is built up of 90 to 95, β(1→4)-linked mannopyranosyl units and nearly all the backbone monomers carry one α(1→6)-linked galactopyranosyl group.

For galactomannans having a lower number of galactose groups than mannose, there exist portions of unsubstituted blocks (the so-called *nonhairy* or *smooth regions*) of mannan backbone, while the grafts are clustered in other (called *hairy*) regions. In the case of locust bean gum, these unsubstituted blocks of the backbone can be as large as 25 mannose units. For guar, these unsubstituted portions are estimated to be approximately <6 mannose units. These smooth regions of the molecular backbone come closer to form interchain hydrogen bonds, which reduce the solubility and hydration of any gum. Thus, the cold-water solubility of LBG is far less (~30%) compared to that of guar gum (~60%–70%). Compared to these, fenugreek gum has >90 cold-water soluble fraction.

Fenugreek gum has very few (~1%) unsubstituted mannoses in the backbone and has no large nonhairy regions on the backbone. Hence, there are practically no chain–chain interactions between the molecules of fenugreek galactomannan to produce insoluble fraction. Cold-water solubility of fenugreek gum is nearly complete (>90%).[20] In a dilute solution it disperses to molecular level and forms no spaghetti-like aggregates.

9.10   CONFORMATION OF FENUGREEK GALACTOMANNAN[21]

A molecule of fenugreek galactomannan in an aqueous solution has a fully extended rod-like conformation, and hence it occupies a large volume of gyration. In contrast to other galactomannan gums, a nearly completely substituted mannan backbone in fenugreek galactomannan has very little chance of interchain association to form spaghetti-like clusters. The hydrated and gyrating molecules of such linear polysaccharides collide with each other and with clusters of water molecules associated to them to produce solutions of high viscosity.

Monosaccharide units in galactomannans possess a *cis*-pair of hydroxyl groups on their pyranosyl ring structure. In the case of mannose, it is the C-2, C-3 hydroxyl pair and the C-3, C-4 hydroxyl pair in the case of galactose that have *cis*-configuration. In contrast to this, galactopyranosyl monomers in cellulose and starch have all *trans*-hydroxyl groups. The linear cellulose molecule has a ribbon-like structure in which rotation around the glycoside bond is restricted. For fenugreek galactomannan, rotation is possible around an interpyranose glycoside bond. The presence of an *cis*-hydroxyl pair along with an essentially linear chain configuration leads to many unique interactions of galactomannans. The hydrogen bonding property of galactomannan polymers toward water is far stronger compared to glucose polymer.

Wide-angle x-ray diffraction pattern of hydrated fenugreek galactomannan shows that it crystallizes into orthorhombic lattice, with $a = 9.12$ Å, $b = 33.35$ Å, and $c = 10.35$ Å, and the density corresponds to four chains passing through a unit cell. This crystal structure is similar to those of other galactomannans.

9.11 PHYSICAL PROPERTIES AND RHEOLOGY OF FENUGREEK GUM[13,14]

Purified fenugreek gum is a nearly colorless or slightly cream-colored powder, which can have a characteristic, faint smell of fenugreek seed. It has a bland taste. Typical composition of commercially marketed, pure and extra pure fenugreek gum powders is given in Table 9.3.

Commercially available fenugreek galactomannan (gum) is easily dispersed in water and hydrates at room temperature to develop high viscosity, in the range 500 to 1500 cps at 1.0% concentration, as measured by the Brookfield viscometer at 25°C at 20 rpm. A sample of this gum prepared by the author using dry processing had a viscosity of ~2000 cps at 1.0% concentration. This is a clear indication of some depolymerization of the product gum made by the water (solvent) extraction process.

Fenugreek gum solutions decrease in viscosity with a rise in temperature and exhibit non-Newtonian, shear thinning viscosity behavior, except at concentrations <0.2%. Its viscosity at a given concentration is much lower than guar gum and locust bean gum due to its lower molecular weight and lack of tendency to form molecular aggregates.

Because of a fully substituted backbone, it does not interact with other polysaccharides (e.g., agar, carrageenan, or xanthan) and does not show synergic viscosity increase and gelling. It is, however, gelled by cross linking due to borate ions, which must be attributed to the interaction of *cis*-hydroxyl groups of different molecular chains.

Nearly complete shielding of mannan backbone by galactose grafts makes the fenugreek polysaccharide backbone hydrophobic, and difficult to be approached and cleaved by β-mannase enzymes, which are normally used for partial depolymerization of other galactomannans. For fenugreek galactomannans with a completely substituted mannan backbone, an initial reaction with α-galactosidase enzymes removes some galactose grafts, exposing bare sites of the mannose chain. This can be followed by an attack by β-mannase enzymes to cleave the polysaccharide, thus causing a decrease in solution viscosity. Controlling of such a sequence of successive enzymes action is, however, difficult.

TABLE 9.3
Percentage Composition of Commercial Fenugreek Galactomannan Powder

Constituent	Pure (%)	Extra Pure (%)
Total carbohydrates	75	85
Proteins	15	5
Lipids (ether extract)	0.4	0.2
Soluble galactomannan	60	75
Ash	1.0	0.5
Water dispersibility	Good	Complete
Smell	Faint	Negligible

9.12 SURFACE ACTIVITY AND EMULSIFYING PROPERTY OF FENUGREEK GUM[13]

Most of the soluble polysaccharides have a certain degree of emulsion-stabilizing action. This has been attributed to high-solution viscosity or the thickening properties of gums, which imparts increased stability to emulsified oil droplets. There is not much reduction of interfacial tension or surface tension in solutions by most galactomannans. In contrast to this, fenugreek gum solution shows the unique property of a large reduction of interfacial and surface tension. This effect is comparable to that of gum arabic, whose action as an emulsifier in solution has been attributed to its composition. Gum arabic is composed of a strongly hydrophilic carbohydrate portion present as a composite with a hydrophobic protein portion and this fulfills the condition for any substance to act as a surface-active agent and to function as an emulsifier.

The mechanism that has been suggested for fenugreek gum acting as an emulsifier is quite different. Highly purified and protein-free (protein <1%) fenugreek polysaccharide is as good an emulsifier as gum arabic. Therefore, it has been suggested that a nearly complete galactose-grafted mannan backbone in fenugreek gum has become hydrophobic, and it has a hydrophilic exterior galactose layer. This allows the fenugreek galactomannan molecule to have a fully extended conformation in solution. Such polysaccharide molecules are deposited on an emulsified oil droplet in water, protecting it against coalescence and flocculation. This emulsifying property of fenugreek gum coupled with its strong moisture-holding capacity opens the interesting possibility of using fenugreek gum in cosmetic products.

9.13 MEDICINAL AND OTHER USES OF FENUGREEK GUM

In India and its neighboring countries, there is a custom that a woman during the postdelivery and lactation period eats fenugreek-based food, because it is thought to promote lactation. In Europe and America, women eat fenugreek seeds as a health food and to enlarge breasts. It is thought that steroidal saponin in fenugreek seed is a precursor of female hormone, which is responsible for enlarged breasts and increased lactation.

Fenugreek seed, having a bitter taste and strongly aromatic odor, has been respected for its various medicinal benefits for over 2500 years. It has been described in ayurvedic and unani systems of health care. Fenugreek has been used as a folk medicine to control blood sugar since ancient times, and more recently it has again attracted the attention of medical personnel for its remedial effect against diabetes.

Some of the medicinal properties of this seed are due to its galactomannan gum, which is a soluble dietary fiber. Being nonmetabolized by the enzymes in the human system, galactomannans act as dietary food fibers and promote beneficial prebiotic colon bacteria. It has been confirmed by animal experiments and clinical tests on human beings that ingesting a food compounded with fenugreek gum lowers the level of blood sugar. It has also been proven that fenugreek seed powder lowers the level of blood cholesterol and other lipids, and it restricts cholesterol biosynthesis in the liver. These effects, that is, sugar and cholesterol control, have been shown to be caused by galactomannan gum contained in albumen in seeds. Fenugreek gum thus acts as a dietary fiber, which has the effect of lowering the level of sugar and cholesterol in blood.

Fenugreek gum, when compared to other dietary fibers, shows maximum efficacy in lowering blood glucose and lipids including low-density lipoprotein (LDL) and cholesterol. In India many people regularly take whole fenugreek seed powder in spite of its bitter taste and odor. In the Western countries, isolated, odorless, and tasteless gum is used as a fiber-rich food additive. Specially made biscuits and beverages containing fenugreek gum are now available in the United States and other Western countries as a dietetic food. Clinical studies have shown that 2 to 3 g/day of fenugreek gum is effective in controlling blood sugar, whereas the requirement of other food fibers is much larger (~20 g).

For controlling diabetes and hypercholesterolemia, it has been recommended to take fenugreek seed powder regularly. However, for some people, particularly in Western countries, it is difficult to take whole seed powder regularly because of its bitter taste and spicy odor. To overcome this problem, manufacturers of fenugreek gum are now producing and marketing its polysaccharide powder, which has no taste or smell, because it is completely free from the husk and germ portions that have bitter and smelly constituents.

A mechanism has been suggested for fenugreek gum to reduce blood sugar and blood lipids. Fenugreek gum thickens food when ingested and forms a gel in the stomach trapping fat, sugars, and starch hydrolyzing amylase enzymes. This results in the slowing of sugar absorption. Thus, it is good for obese and diabetic persons. The gel, which appears like fat inside the body, signals the brain to send a message to the gall bladder to empty bile in the stomach. The gel then irreversibly traps lipid-emulsifying bile salts and prevents their reabsorption. Thus, emulsification and absorption of lipids including cholesterol results in the lowering of blood lipid. This in turn reduces hypertension and chance of heart attack.

9.14 CONCLUSION

Commercial production of fenugreek gum is still a new venture in India. It is very likely that its use in specialty food and beverages should increase. It has been stressed upon earlier that success to keep its production cost low should depend on adopting a dry grinding process for its extraction. Fenugreek seed is already being produced in India and its agriculture can be increased, as has been done in case of guar. It is quite certain that fenugreek gum is not likely to have many nonfood uses, just like guar, but it has already emerged as a useful food additive. In western Rajasthan in India there is a well-established guar gum industry. The technology base of the guar gum industry can be well extended to produce fenugreek gum and this can make India a pioneer for production of yet another galactomannan polysaccharide.

REFERENCES

1. New Vistas in Pulse Production. Indian Agricultural Research Institute, New Delhi, 1971.
2. Arora, S.K., Ed., Chemistry and Biochemistry of Legumes, Oxford & IBH Publishing Co., New Delhi, 1982.
3. Cui, S. W., Polysaccharide Gums from Agricultural Products: Processing, Structure, and Functionality, Technomic, Lancaster, PA, 2001.
4. Whistler, R. L. and Hymowitz, T., Guar: Agronomy, Production, Industrial Use, and Nutrition, Purdue University Press, West Lafayette, IN, 1979.
5. Anonymous, Tropical Legumes: Resources for the Future, National Academy of Sciences, Washington, D.C., 1979.
6. Intellectual Property Rights Bulletin, Vol. 9, No. 11, November 2003. National Botanical Research Institute, Lucknow (CSIR, India).
7. Whistler, R. L. and BeMiller, J. N., eds., Industrial Gums, 3rd ed., Academic Press, New York, 1973.
8. Reid, J. S. G., Galactomannans from legumes seeds endosperm, Adv. Bot. Res., 1 (1985): 125–155.
9. Zimmer, R. C., Can. Plants Dis. Survey, 64 (1984): 33.
10. Cui, S. W. Polysaccharide Gums from Agricultural products: Processing, Structure and functions., Technomic, Lancaster, PA, 2001, 238–247.
11. Mathur , V. and N. K. Mathur, J. Sci. Ind. Res., 64 (2005): 475–481.
12. Meier, M. and Grant, J. C., Planta, 133 (1977): 242–246.
13. Garti, N., Madar, Z., Aserin, A., and Sternheim, B., Food Sci. Tech., 30 (1997): 305–308. Garti, N., Food Hydrocolloids, 8 (1994): 155.
14. Mathur, N. K. and Mathur, G. M., Fenugreek gum, Science-Tech Entrepreneur, December 2006.
15. Reid, J. S. G. and Meier, H. Z., Phytochem., 9 (1970): 513.
16. Iyer, C. R. H. and Sastri, B. N., J. Indian Inst. Sci., 16A (1955): 88.
17. Hui, P. A. and NeuKom, H., Tappi, 47 (1964): 39.
18. Garti, N. and Pinthus, E. J., NutraCos, 1 (2002): 5–10.
19. McCleary, B. V., Carbohyd. Res., 71 (1979): 205.
20. Mathur, N. K., Mathur, V., and Nagori, B. P., In Trends in Carbohydrate Chemistry, Vol. 3, P. L. Soni, Ed., Surya Publishing, Dehradun, India, pp. 19–123.
21. New Vistas in Pulse Production, Indian Agricultural Research Institute, New Delhi, 1971.

10 Tara Gum
An Exclusive Food Additive with Limited Production

10.1 INTRODUCTION

The plant gums industry has been looking for newer sources of polysaccharide gums, particularly the galactomannans.[1] The search for less explored plants as sources of gums becomes more important when the current source of a particular gum with certain specific functional properties has insufficient production to meet the increasing market demands.

Insufficient availability of a gum can result in frequent escalation in price of that particular gum that is currently being used as an additive for food. This has been the case with locust bean gum (LBG; Chapter 8). Currently the worldwide production of LBG is around 20,000 tons per annum, whereas its demand is more and increasing regularly. This has resulted in an ever-increasing price of LBG, and a search for alternative substitutes. The need for a galactomannan gum with specific properties similar to LBG as an additive for food has resulted in developing tara gum as a food hydrocolloid.[2]

Limitations in increasing the production of LBG have been discussed earlier (Chapter 6). To a certain extent, these limitations can be met by exploring other galactomannan gums having a lower percentage of galactose. Considering the fact that total production of LBG is likely to remain stagnant at the present level, attempts are ongoing to find its substitute in other tree gums. Attempts to modify functional properties of certain abundant polysaccharides (e.g., guar gum) to replace LBG have met with very limited success.

If we look at the history of the introduction of guar gum into the U.S. paper industry to substitute for LBG, we find that the use of guar gum in certain industrial fields, for example, oil-well drilling, food additives, and textile printing, increased rapidly. Guar gum consumption also got a boost due to the introduction of its modified and derivatized products, which have found numerous industrial applications. Still guar gum has not been found to be a suitable substitute for LBG, at least in specific food products (i.e., food gels, ice creams, and other frozen foods). Hence, there has been an increasing demand for LBG or its equivalent gums in the food industry, where galactomannans having lower galactose content (mannose-to-galactose ratio [M:G] of 3:1 or less) can only be used.

With a lower galactose content, any galactomannan should have a lower water solubility and it is likely to exhibit a mixed sol-gel or semigel rheology (for details, see Chapter 8). Due to the regional and climatic requirements, the production of

LBG is not likely to increase to meet increasing demand. Substitute gums for LBG from other plant sources are now being explored. One such substitute has been found in tara gum.[3]

Generally the galactomannans from full-grown and perennial trees, rather than from the annual plants (guar and fenugreek), have lower galactose content. The search is on for legume trees producing such galactomannans. LBG has not only the optimum M:G ratio, but also a favorable placement of galactose stubs on the mannan backbone, which makes it suitable for use in food. The use of LBG-type gums is not likely to diminish except for price consideration and that too with some loss in its effectiveness as a food additive.

10.2 TARA TREE

With approximately 650 genera and 18,000 species, Leguminoseae is one of the largest families of flowering and pod-bearing plants. Plants of this family are found worldwide, including in the temperate zones, humid tropic regions, semiarid zones, savannas, and low lands. Among the full-grown trees, the acacias, carob, *cassia fistula* and tara plants are easy to recognize.

The tara shrub is a small leguminous tree or thorny shrub.[2] Its botanical name is *Caesalpinia spinosa* and is commonly known as molina kuntze or the tara tree in Peru. Some other names for the tara shrub are *Poinciana spinosa* Mol., *Caesalpinia pectinata* Cav., *C. tara*, *C. tinctoria* HBK, *Coulteria tinctoria* HBK, *Tara spinosa*, and *Tara tinctoria*. Its common names are spiny holdback, tara, taya, and algarroba tanino.[3] The tara tree, which grows in the Andes region of Peru, is also referred as *harangue, guaranga,* or Peruvian carob.

The tara shrub is native to Peru and Peru continues to be the major producer of tara gum. The tara shrub can also be found growing, but to a far lesser extent, throughout the northern, western, and southern parts of South America (i.e., from Venezuela to Argentina). It has now been introduced into the semiarid parts of Asia, the Middle East, and Africa and has become naturalized in the state of California in North America. The tara shrub has been mainly cultivated as a source of tannins, used for the specialty leather industry.[4,5]

Tara is a pod-bearing legume tree and besides tannins, it is also a source of unique galactomannan gum. The tree has also been grown as an ornamental plant because of its large and colorful flowers and pods. The tara tree typically grows 2–5 meters tall. Its bark is dark gray with scattered prickles and hairy twigs. Tara leaves are alternate, evergreen, lacking stipules, bipinnate, and lacking petiolar and rachis glands. Leaves consist of 3–10 pairs of primary leaflets under 8 cm in length, and 5–7 pairs of subsessile elliptic secondary leaflets, each about 1.5–4 cm long. Inflorescences are 15–20 cm long terminal racemes. Many of these trees are flowered and covered in tiny hairs. Tara flowers are yellow to orange in color, with 6–7 mm petals. The lowest sepal is boat shaped with many long marginal teeth. Its stamens are yellow, irregular in length, and barely protruding (Figure 10.1).

The tara tree bears fruits, which are flat, oblong, and indehiscent pods. These pods are about 6–12 cm long and 2.5 cm wide, and each pod contains 4–7 rounded black seeds, which become reddish in color when they are matured. The seed of the

FIGURE 10.1   Flowering tara tree.

tree consists of about 38% hull, 40% germ, and a rather low 22% of endosperm. The endosperm of this seed is mainly composed of a galactomannan polysaccharide.

10.3   TARA GUM

It has been mentioned earlier that plants from which galactomannan-based gums are obtained, grow in many areas, which are scattered all over the world, including South America. Tara gum is a unique galactomannan polysaccharide, which is obtained from the seed of the legume tara tree.[3] The M:G ratio in tara gum is approximately 3:1, that is, it is intermediate between guar gum (2:1) and LBG (4:1). In physical and functional properties, tara gum behaves like a blend of guar gum and LBG. Tara gum is produced in much smaller quantities of only about 1000 tons per annum, compared to the two other major galactomannans (guar gum ~100,000 and LBG ~20,000 tons per annum). Due to it's being produced in much smaller amounts, so far tara gum has not found its niche in the world of commercial hydrocolloids or gums. Still there is a possibility that its blends with other gums, such as xanthan, which has a unique rheology and can open new vistas for its use in food.

Tara gum is white or beige in color, and a nearly odorless powder that is produced by separating and grinding the endosperm of *Caesalpinia spinosa* seeds. The processing of tara gum is similar to the manufacturing of LBG. The major component (~80%) of tara seed endosperm powder is a galactomannan polymer.[6] Tara gum in its chemical structure is somewhat similar to the main polysaccharide components of guar and locust bean seeds. Gums from the latter two plants are also used widely in the food industry.

Tara gum has been deemed safe for human consumption and as a food additive, and it is currently being used as a thickener and stabilizer in a number of food applications. Being of lower molecular weight tara galactomannan produces a far lower viscosity (300–400 cps at 1%) in aqueous solution when compared to an aqueous

TABLE 10.1
Composition of Typical Sample of Tara Gum

Component	Percentage
Moisture	12–15
Lipids	0.2–0.5
Protein	2–3
Cellulosic fibers (insolubles)	1.0–1.5
Ash (minerals)	2–3
Galactomannan (by difference)	80–85

guar gum solution (4000–7000 cps) at the same concentration, but it forms a more viscous solution than that of locust bean gum. Blends of tara gum with modified and unmodified starches have been produced, which have enhanced stabilization and emulsification properties, and these are now being used in the preparation of frozen foods (e.g., ice cream).

Just as other seed gums, tara gum is also produced by separating and grinding the endosperm portion of the *Caesalpinia spinosa* seed, and it is a whitish and almost odorless powder. The germ portion of the seed is rich in protein, while the hull is mainly composed of cellulose, containing tannins. The major constituent of tara gum is a galactomannan polysaccharide, which is similar in its chemical structure to other commercial galactomannan gums (e.g., guar gum and LBG). Typical composition of commercially sold tara gum is shown in Table 10.1.

10.4   STRUCTURAL FEATURES OF TARA GUM

Preliminary structural studies of tara gum have shown it to have the usual structural features of legume galactomannans, that is, it has a $\beta(1\rightarrow4)$-linked mannan backbone having randomly substituted, $\alpha(1\rightarrow6)$-linked galactose grafts.[7] The distribution of galactose grafts on mannose backbone in tara gum has not been studied in complete detail, but its binary gelling of xanthan and agar polysaccharides shows that galactose grafts are likely to be present in continuous blocks (hairy regions), separated by blocks of bare or nonhairy regions, somewhat similar to that in LBG. Both LBG and tara gum upon cooling undergo self-gelling to form weak gels, and hence these can be used only in mixed gel systems. A further comparison of the properties of LBG and tara gum can be made. Thus, 60% sugar concentration 0.2% LBG forms a weak gel, whereas tara gum forms a comparable gel at 0.5% concentration only. In self-gelling behavior, there is another $\beta(1\rightarrow4)$-linked polysaccharide, konjac glucomannan. Though weakly self-gelling, konjac polysaccharide is also used in mixed gels only.

Tara gum shows the usual chemical properties of galactomannans, for example, it forms a gel with borax and transition metal ions. It shows a pseudoplastic rheology, and its viscosity decreases with increasing shear rate and with a rise in solution temperature.

10.5   MANUFACTURE OF TARA GUM[8]

Just like locust bean gum, the processing of Tara seed into its gum powder is bestowed with problems of less effective dehusking of the seed and difficulty in powdering of its very hard endosperm. As a generalization, galactomannans from full-grown trees have lower galactose content and their seeds are larger and harder as compared to those from annual crops (guar and fenugreek).

The hardness of tara seeds and other galactomannan-bearing tree seeds can be attributed to the distribution of galactose grafts on its mannan chain in large continuous blocks separated by bare portions of the mannan backbone. Such molecular structure of a linear polysaccharide can cause strong chain–chain interactions, resulting in a dense packing of the polysaccharide molecules in the seed endosperm, producing hardness.

Additionally, the gum content of tara seed is low. Still tara gum got commercialized because its pod being rich in pyrogallol tannins was already being harvested. Tara tannins find application in tanning of specialty leather. After recovery of tannins, by powdering dried pod-skin, the kernel (seed) is available for the extraction of gum.

Current manufacturers of tara gum are Cesalpinia S.p.A. and Nutragum S.p.A. in Italy and Unipectin AG in Switzerland. These companies have to import the tara kernel from Peru, where there are no facilities for extraction of its gum. In 1981, these hydrocolloid companies applied for and got approval for tara gum as a food additive from the Food and Agricultural Organization (FAO) of the United Nations Organization (UNO). Currently the acceptable daily intake (ADI) of tara gum has been limited to 12.5 mg/kg of body weight. More toxicological information about the gum is being collected to get unrestricted approval for tara gum in food. The composition of food-grade tara gum was shown earlier in Table 10.1.

10.6   APPLICATIONS OF TARA GUM[9,10]

Tara gum shows a pseudoplastic rheology. Its viscosity decreases with increasing shear rate and with a rise in the temperature of its sol. It is currently produced in a small amount, which is solely used in the food industry as a thickener and stabilizer. When used in dessert gels, seafood, meat, ice cream, and yogurt it gives an improved rheology, better water binding, and emulsification effect.

Patents related to tara gum claim it to be a water-soluble polymer, which has been cited as having many applications, comparable to those of guar or LBG. Economic factors limit the use of tara gum, its low production amount (~1000 tons annually), and lack of a plentiful and reliable source of the gum.

10.7   CAN INDIA DEVELOP AN EQUIVALENT OF
TARA GUM FROM *CASSIA FISTULA* GUM?

Cassia fistula (*amaltas* in Hindi) produces a galactomannan gum, which is present in the endosperm of its seeds obtained from its pods. This is one of the far less investigated legume trees of Indian origin, which can become another industrial source of a food galactomannan gum. *Cassia fistula* can emerge as yet another promising galactomannan gum source in India, which might prove to be an equivalent of tara gum and a suitable substitute of LBG.

Amaltas is a decorative tree that grows all over India and Southeast Asia. The tree can be observed in full bloom during the peak summer. During its period of maximum flowering, which is the midsummer season (April–June), the tree appears to have more flowers (yellow) than leaves (green), with the result that the whole tree appears to be bright yellow in color. It bears long, green pods (20–30 cm) that become dark brown or almost black upon ripening (December–January). These, long dark pods are seen hanging on the trees. Dry and ripened pods tend to fall away from the tree. The seeds, now detached from the dried pulp inside the pod, rattle on shaking the pod. When the pod is broken, the seeds come out. Though it has not so far been investigated, the outer bark of its pod is likely to be a rich source of tannins, just like that of the tara pod.

The gum from the seeds of *Cassia fistula* has been investigated by Dr. V. P. Kapoor, a scientist (now retired) at the National Botanical Research Institute at Lucknow, India, who has found it to be a galactomannan having its M:G ratio very close to that of tara gum. Looking to the requirements for a LBG substitute, there is a need to make toxicity studies and to develop a technology based on dry grinding for extraction of endospermic gum from the *Cassia fistula* seed, which might prove to be a good substitute of LBG. This can easily be done by following the technology of dry processing of guar seed, now being widely practiced in India.

Just like in Peru and many other developing countries of South America, amaltas can become a livelihood and life support for some tribal poor in India. It is likely that the total quantity of gum produced from this tree may not be as much as that of guar gum or LBG, but as a combined source of tannin and a food gum, it shall be worthwhile to work for economical exploitation of this regenerable natural resource.

REFERENCES

1. Duke, A., Ed., Handbook of Legumes of World Economic Importance, Plenum Press, New York, 1981.
2. Food and Agriculture Organization of UNO, Rome, Specifications for Identity and Purity, Food and Nutrition Paper 4, 1978, pp. 52–53.
3. Anderson, E., Ind. Eng. Chem., 42 (1949): 288.
4. Rogers, J. S. and Beebe C. W., J. Am. Leather Chem. Assoc., 36 (1949): 525.
5. Garro Galvez, J. M., Riedl, B., and Conner, A. H., Analytical studies on Tara tannins, Holzforschung, 51 (1997): 235–243.
6. Borzelleca, J. F., Ladu, B. N., Senti, F. R. and, Egle, J. L. Jr. Evaluation of the Safety of Tara Gum as a Food Ingredient: A Review of the Literature, J. Am. Coll. Toxicol., 12 (1993): 81–89.
7. Buffington, L. A., Stevens, E. S., Morris, E. R., and Rees, D. A., Int. J. Biol. Macromol., 2 (1980): 199.
8. Moe, O. A., Miller, S. E., and Buckley, M. I. J. Am. Chem. Soc., 74 (1952): 1325.
9. Dea, I. C .M., Morris, E. R., Rees, D. A. Eelsh, E.J. Barnees, H. A. and Price, J., Carbohyd. Res., 57 (1977): 249.
10. Dea, I. C. M., Interaction of ordered polysaccharides. Structure-synergism and freeze-thaw phenomenon, Proc. Easter School Agr. Sci., University of Nottingham, (1979): 229–247.

11 Cassia tora Gum
An Emerging Commercial Galactomannan

11.1 INTRODUCTION

Some polysaccharides, which are hydrocolloid by nature, are extracted from the seed endosperms of wild legume plants. Polysaccharide gums thus obtained and when found nontoxic can generally be used as thickening and gelling agents in food and in several nonfood industries. Galactomannan gums are present as the endospermic reserve polysaccharides in a large number of legume seeds.[1] Many of the annual agricultural crops, wild annual herbs, and full-grown perennial trees and shrubs have galactomannans in their seeds.[2] Thus, after examining hundreds of legume seeds, E. Anderson, concluded that galactomannan polysaccharides are a common constituent of many legume endosperms.[3,4] Similar conclusions were also drawn by Dr. V. P. Kapoor, emeritus scientist of the National Botanical Research Institute in Lucknow, India, who has made an extensive survey of the *Cassia* species of plants and has found them to be a rich source of galactomannans. According to his survey, there are still many, unexplored *Cassia* plants indigenous to India that could become a future source of industrial gums.[5]

From economic considerations and availability of plant seeds, which are the raw materials for gum extraction, the current commercially produced galactomannan polysaccharides come from relatively few endospermic material of leguminous plant seeds. Commercial galactomannan gums, which are now produced are those from the plants *Cyamopsis tetragonoloba* (guar gum), *Caesalpinia spinosa* (tara gum), *Ceratonia siliqua* (locust bean gum [LBG] or carob gum), *Trigonella foenum-graecum* (fenugreek gum), *Sesbania bispinosa* (daincha gum), and *Cassia tora* (CT gum).

11.2 COMMERCIALLY PRODUCED GALACTOMANNAN GUMS[6]

No doubt guar gum is the largest produced galactomannan gum, with a production of more than 100,000 metric tons per year, while locust bean gum at ~20,000 Mt per year is the second largest produced galactomannan. Both these gums are approved for use in food. Tara gum has been commercially produced since the year 1981 and fenugreek gum production commenced in the year 2000, and its production has steadily increased. These two gums are more recent additions to the list of commercially produced galactomannans. Both of these gums have been approved for use in food. *Sesbania bispinosa* gum or daincha gum, and *Cassia tora* gum (CT gum) have been produced from wild, annual herbs since the late 1960s, but these were used mainly for nonfood, commercial uses. Frequently these gums are used as a cheaper substitute for guar gum in nonfood industrial applications.

Cassia tora gum is currently being used mainly in pet food as a binder. There is, however, a strong case being made to obtain permission for approval of this gum as a human food additive in the United States and European Union (EU) countries. Such approval for human food applications of *Cassia tora* gum has already been obtained in France, Belgium, and Austria. More complete EU approval was expected by the end of 2009. Approval in the United States is still pending (2009) with no clear indication of when it shall be granted. More recently purified CT gum has acquired a status, very close to GRAS (generally regarded as safe as a food additive) specification in the United States.

Collection of *Cassia tora* seeds and its processing into CT gum in India is currently confined to an unorganized sector consisting of small-scale industries. India is also a major CT gum-producing country. Looking to its potential future use in the food industry, there is a need for more research and development efforts in India, particularly on its purification for the food industry and its derivatization to increase its aqueous solubility for use in the textile industry.

11.3   CHEMICAL STRUCTURE OF *CASSIA TORA* GUM[7]

Molecules of *Cassia tora* galactomannan gum are composed of a linear backbone of $\beta(1\rightarrow4)$-linked D-mannopyranosyl units with recurring, single, $\alpha(1\rightarrow6)$-linked D-galactopyranosyl groups, randomly branching from C-6 of mannopyranose units in the backbone. This is the usual structural feature of legume galactomannans. Though galactomannans from different legume plant species have this common structural feature, they differ in their molecular weight, frequency of the occurrence of the galactosyl side chains (i.e., their mannose-to-galactose ratio [M:G] ratios) and the mode of the placement of galactose grafts on the mannan backbone. The average ratio of mannose and galactose units in the galactomannan from *Cassia tora* seed gum is reported to be approximately 5:1. This means that this gum has one of the lowest percentages of galactose among the galactomannans. This M:G ratio is in a range that is a desirable characteristic of galactomannans needed as food additives.

In Figure 11.1, an idealized molecular fragment of CT gum having an M:G of 5:1 is shown. This however is not a true picture, because the galactosyl side chains are not placed in a regularly repeating pattern, but rather randomly and in continuous blocks, which are separated by bare regions of the backbone.

-β-D-Man-(1→4)-β-D-Man-(1→4)-β-D-Man-(1→4)-β-D-Man--(1→4)-β-D-Man

↑

(1→6)-α-D- Gal

FIGURE 11.1   An idealized fragment of CT gum molecule having M:G = 5:1.

Incomplete water solubility of this gum and its self-gelling behavior suggest that the galactose grafts on the mannan backbone of the molecule are arranged in large continuous blocks of grafts separated by even larger bare regions, similar to those in LBG molecules.

11.4 DEPENDENCE OF THE FUNCTIONAL PROPERTIES OF GALACTOMANNANS ON THEIR M:G RATIO[7]

To a certain extent the presence of galactose grafts on the linear polymeric mannan backbone restricts chain–chain interactions between galactomannan molecules. To a variable extent, these galactose grafts also induce water solubility or dispensability to these gums. Water solubility in a galactomannan arises due to limiting of mannan chain–chain interactions, which can otherwise result in insoluble, spaghetti-type molecular aggregation of the linear molecules. When the percentage of galactose in a galactomannan is low (unlike that in guar and fenugreek gums), resulting in a fewer number of short grafts, it can still result in the formation of more durable chain–chain junction zones. This happens when 10 or more unsubstituted mannose monomer graft units on the polymer backbone are present in a continuous row.

Due to the presence of such unsubstituted backbone regions, which are present in both LBG and CT gum, there is a strong chain–chain interaction between the molecules of these gums, which form molecular aggregates rendering them only partially water soluble or dispersible. Hence, these gums are said to have a mixed sol-gel rheology. This is also true of many other, less investigated galactomannan gums, which have a lower percentage of galactose (M:G = 3:1 or less).

11.5 GELLING GUMS[8,9]

Purified cassia gum, which is obtained from the seed endosperm of *Senna obtusifolia* (also called *Cassia obtusifolia*) and *Cassia tora*, is now regarded as a food additive, thickener, and gelling agent for food products. Purified CT gum, now being used as an additive for food, it has been assigned E number E499 by the European Union.

Cassia tora gum has weak, natural gelling properties. Hence, additives for food gelling composition and thickener can be based on the purer quality of *Cassia tora* gum. Cassia gum has also been used for making air-freshener gel compositions in the form of water gels.

Many hydrocolloids or their mixtures are capable of forming aqueous gels of different gel strength depending on their structure and concentration in the gel in addition to other environmental factors such as the ionic strength, pH, and activity of water and temperature of the gelled material.

The mixed sol-gel behavior (viscoelasticity) of an aqueous hydrocolloid system can be examined by determining the effect that an oscillating force has on the movement of the gelled material. With viscoelastic hydrocolloids, some of the deformation caused by shear stress is elastic, which results due to contortion (twisting) of polymer chains into high energy conformations. These polymer chains again return to the lower energy conformation when the applied force (stress) is removed. The remaining deformation, which is due to the sliding displacement of the chains through the solvent, does not return to the lower energy level when the force is removed. Thus, under a constant force the elastic displacement remains constant, whereas the sliding displacement continues to increase regularly.

Gums, which are linear polymers and have tertiary and quaternary structures under certain conditions, can form aqueous gels. In these gels a large amount of water gets imbibed into a highly cross-inked structure of linear, macromolecular chains of the hydrocolloids. Junction zones between the polymer chains formed in such cases are strong enough to prelude complete hydration of the gum to make them dissolve in water at the molecular level. Molecules of certain gelling polysaccharides, typically carrageenan, agar, and xanthan, interact with LBG and certain other gums to form firmer, binary gels. In presence of LBG-like polysaccharides, the gelling gums can also form a gel at their otherwise nongelling concentrations. For economical considerations, formation of these binary gels has found industrial applications for making frozen and gelled foods.

Galactomannans by themselves do not have any tertiary structure and do not have the tendency to form a firm gel, but they do enhance the gel strength of binary gels. The binary gelling property is associated with those galactomannans that have M:G ratios 3:1 or less.

Such galactomannans are in great demand in the food industry and efforts are being made to find newer, edible sources of binary gelling galactomannans. Until 1990, LBG was the only polysaccharide for use in binary gelling systems available for the food industry. Now tara gum (M:G =3:1) has also been used as a substitute for LBG (M:G = 4:1). As a measure to combat the short supply of LBG, its nonfood uses have been reduced and this gum is now almost exclusively being used as a food additive. Also there has been a large price escalation of LBG, which has nearly doubled to $10/Kg (Rs. 400–450/Kg) during the past two decades.

11.6   MECHANISM OF BINARY GELLING[10]

Traditionally, addition of LBG at low concentration has been used to reinforce the junction zones, by interaction with the helical structure of a gelling polysaccharide in a water medium. The long, bare regions of the mannan backbone of one LBG-type molecular chain can interact with several helices to reinforce and stabilize a three-dimensional network with liquid imbibed into the reorganized portions. Such a structure can hydrate and immobilize the fluid into a gel. *Cassia tora* gum (M:G = ~5:1) has lower galactose content than LBG and has even better binary-gelling properties compared to LBG.

CT gum, originating from an annual herb, can be a more abundant, cheaper, and good substitute for binary gelling systems in which LBG is currently used. There should soon be a need for a purified and edible grade CT gum. Hence, attempts are now being made to completely eliminate or to reduce the toxic constituents in CT gum to make acceptable for its use in human food. Purified CT gum has already been approved for food use in some countries of Europe and it is listed as a stabilizer, thickening, and gelling agent for the manufacturing of canned pet foods. The available data on the quality of purified CT gum demonstrates a lack of any toxic effect in animals and it has been approved in Japan as a human food additive.

11.7   COMMON TOXIC CONSTITUENTS IN LEGUME SEEDS

Though rich in proteins and galactomannans polysaccharides, many legume seeds have been reported to contain certain toxic constituents. These include protease inhibitors and hemagglutinins or lactins, which are proteins by their chemical nature.[11] There can also be many nonproteins and nonpolymeric compounds as toxic constituents in the commercially isolated legume galactomannans. These include cyanogens, saponins, and anthraquinone compounds, which are pigments.

In the case of some wild-growing annual plants (e.g., *Cassia tora*), some of these toxic constituents may have naturally evolved as deterrents against animals grazing on them as food. Most of the agricultured seed grain crops now being used in human food were developed by long periods of their domestication and selection from many wild plants (grasses and legumes) over thousands of years. On selective breeding over long periods, the toxic constituents in these plants were gradually reduced or were completely eliminated. Many wild plants still have these in quantities that can be toxic to humans. Of these toxic constituents, those that are proteins by a chemical nature (e.g., protease inhibitors and lectins) are easily denatured and rendered harmless during the normal cooking of food. Thus, the soybean is a classical example of a legume seed containing both protease inhibitor and lectins, which are rendered harmless in cooking, but nonpolymeric and low molecular weight substances (e.g., toxic anthraquinone compounds) may still remain unchanged.

11.8   ANTHRAQUINONE PIGMENTS IN LEGUME SEEDS

There has been some concern about possible mutagenic effects of a related family of anthraquinone compounds, which includes chrysophanol (also known as rheic acid or chrysophanic acid). Chemically it is, 1, 8-dihydroxy-3-methylanthraquinone (CAS No. 481-74-3).

Similar anthraquinone pigments are also present in aloe vera leaves. The U.S. chemical hazard rating number label for aloe emodin, emodin, and chrysarobin, which are some closely related anthraquinone compounds, present in aloe vera gel, are

National Fire Protection Association (NFPA) health hazard = 0
Flammability = 0
Reactivity = 0

These compounds have been evaluated, but data on human toxicity have so far been contradictory.

In the year 2002 the U.S. Food and Drug Administration (FDA) ruled that certain "stimulant-laxative products" containing chrysophanol be considered as non-GRAS. However, many such products are in common use in medicinal formulation, and chrysophanol itself does not appear to pose a major human health threat.

As an additive in food, the amount of any galactomannan used is normally <1.0% of solids, while the so-called toxic constituents are present in the gums at a parts per million level. Hence, the net effective concentration of such compounds is likely to remain far below any unsafe and toxic limit for human food.

Anthraquinone pigments are intensely colored. One limiting factor in use of *Cassia tora* and similar gums can be the color it imparts on the food product. Even at very low concentration, anthraquinone pigments exhibit an intense brown-red color. Washing of the *Cassia tora* gum, after its extraction from seeds, with a 1.0% alcoholic caustic, followed by neutralization and final aqueous alcoholic wash can completely remove the brownish-red color. Color removal can also serve as a simple assurance for removal of most of the other toxic constituents. The cost and the labor involved in such treatment may be worth doing, considering the fact that the product, so treated, is likely to substitute for LBG.

11.9   *CASSIA TORA* CROP[12]

Cassia tora belongs to the Leguminosae family, and it grows in dry soils and in hot and wet tropical and subtropical climates. It is also occasionally cultivated for its gum. As a wild weed or a cultivated crop, *Cassia tora* grows in most parts of India. It has been considered as a legume crop for improvement of wastelands. Looking to the emerging economic aspect of CT gum and the plant as a fodder crop, Australia is considering introducing it as a dual-use crop, particularly in its northern wastelands.

Cassia tora gum is derived from the plant *Cassia tora*, which is a wild weed in the Caesalpinaceae family. Its common names are sickle pod, sickle senna, fortid cassia, tovara, coffee pod, and chakvad. Cassia is basically a slim, evergreen plant. In order to harvest seeds of this plant, it is now grown as a cassia crop, which results in the improvement of soil.

11.10   *CASSIA TORA* SEED

A typical cassia seed powder contains 1%–2% volatile cassia oil, which is mainly responsible for its spicy aroma and taste. CT gum is a natural gelling agent that has many industrial uses. It can also have applications as a food additive. CT gum is commercially extracted from *Cassia tora* seed. The whole cassia seed powder is known to be a tonic, carminative, and stimulant. In India it is also used as a natural pesticide in organic farming. Primary chemical constituents of cassia seed include cinnamaldehyde, galactomannan polysaccharide, tannins, mannitol, coumarins, and essential oils (aldehydes, eugenol, and pinene). Roasted seeds are substituted for coffee, like tephrosia seeds. It also contains sugars, resins, and mucilage among other constituents. Since *Cassia tora* continues to grow in the wild, the fitness of its gum for food long remained questionable and its use is mostly confined to technical applications (e.g., textile printing). It was used where its functional properties permitted its use as a substitute of guar gum. Guar gum, being abundant and cheap, has almost completely substituted CT gum for its technical applications.

11.11   *CASSIA TORA* GUM AND ITS STRUCTURE[13]

The technology of manufacturing CT gum has essentially been adopted from those for making LBG and guar gum. Because of the greater hardness of CT split compared to that of guar, grinding is difficult. The gum produced can have more impurities of proteins and acid insoluble residue (cellulose). CT gum is generally marketed as much coarser powder (80–120 mesh) compared to guar gum (200–300 mesh).

Cassia gum is the purified endosperm powder of the seeds of *Cassia tora* and *Cassia obtusifolia*. The gum is made from *Cassia tora* seed splits by grinding in a way that is similar to that of guar gum. Commercial CT gum is generally comprised of at least 75% galactomannan polysaccharide consisting primarily of a linear chain of β(1→4)-linked, D-mannopyranose units with α(1→6)-linked single D-galactopyranose branches or grafts. It is a high molecular weight galactomannan polysaccharide, which is a hydrocolloid by nature. CT gum is made in a technical grade and semirefined grade. Commercial gum is the purified powder of the endosperm of the seeds of *Cassia tora/obtusifolia,* which belong to the Leguminosae family.

Some specifications of commercially available CT gum of 60–100 mesh (standard U.S. mesh) powder are given in Table 11.1.

Cassia gum or the endosperm powder of *Cassia tora* seed split is well suited for producing gels in combination with other gums and thus it can have potentially numerous food applications. It is currently used as a thickening and binding agent in pet food only, but it can have other food applications, for example, emulsification, foam stabilization, moisture retention, and texture improvement, similar to other edible gums. Composition and some characteristics of CT gum are given in Table 11.2.

TABLE 11.1
General Specifications of Technical and Refined-Grade CT Gum

Hydrocolloid (CT Gum)	Technical Grade	Refined Grade
Appearance, particle size	Yellowish, 60-mesh powder	Whitish, 100-mesh powder
Source	*Cassia tora* endosperm	Refined *Cassia tora* endosperm
Chemical nature	Galactomannan	Galactomannan (refined)

TABLE 11.2
Composition and Properties of CT Gum

Major Constituents	Percentage Composition
Cellulosic fiber	5–12
Galactomannan gum	70–80
Protein	4–8
Ash	1.5–2.5
Acid insoluble residue	7–13
Lipids (fat and volatile oil)	0.5–1.5
pH 5% aqueous dispersion	6–8
Viscosity, 5% sol, made in hot water and cooled to 25°C (Brookfield viscometer) spindle no. 6, at 20 rpm	15,000–50,000 cps

11.12 RHEOLOGY OF *CASSIA TORA* GUM[14,15]

CT gum is insoluble in water at room temperature; it only swells in it. In hot water (>80°C) it forms a viscous, colloidal sol in which the insoluble portion of the gum remains suspended as a gel. It has a good thickening ability for food when used as a single ingredient and more so in synergy with xanthan (microbial gum) or carrageenan (seaweed gum). Because of its synergistic properties, it forms firm, thermoplastic gels with carrageenan and xanthan, which have improved stability and water retention.

It has been suggested that the gelling polysaccharides (e.g., carrageenan, xanthan, and agar) have a partially organized, tertiary or helical structure in solution, even at nongelling concentrations. When the temperature is lowered the molecular chains reorganize further into a three-dimensional structure due to the formation of durable junction zones resulting in gelling.

11.13 GRAS SPECIFICATION AND E NUMBERS FOR FOOD ADDITIVES

In the United States, all food additives are subject to premarket approval by the Food and Drug Administration (FDA), which means its use is generally recognized as safe (GRAS), as certified by qualified experts. In 2003 and 2004 the FDA received GRAS notices for CT gum on behalf of food-grade CT gum manufacturer Noveon Inc. The notices are still being considered by the FDA. In June 2008 the U.S. firm Lubrizol Advanced Material filed a petition to the FDA proposing that food regulations be amended to provide for the safe use of cassia gum as a stabilizer in frozen dairy desserts.

Just like the GRAS specifications, the addition of E-numbered additives in food products has been an issue of ongoing health concern in Europe. Some of the food additives may be of genetically modified origin, which are as yet not permitted in human food in Europe. E number codes for food additives are usually found on food labels throughout the countries of the European Union. E numbers are also encountered on food labeling in other jurisdictions, including Australia. Food-grade CT gum has been assigned the E number E499.

11.14 PRESENT STATUS OF CT GUM

CT gum swells in cold water and it forms a high-viscosity aqueous colloidal sol when it is boiled in water. Purified gum is well suited for use in the manufacture of food gels in combination of other gums and it can have potentially numerous other food applications. It is used as a thickening and binding agent in pet food. Its other applications include emulsification, foam stabilization, moisture retention, and texture improvement agent when used at concentrations comparable to those of other edible natural gums. Use of CT gum powder in the pet food industry is becoming more popular.

Extensive studies are now being carried out on CT gum to extend its applications in human food. Diamalt AG (a branch of B.F. Goodrich, Germany) is currently manufacturing CT gum in India, according to their patented process, in which the amount of chrysophanic acid is very much reduced and it is suitable as a binder for pet food. This patent has now expired, which allows other companies in India to manufacturer CT gum and to further improve it as a food-grade gum additive. It was mentioned that CT gum is approved for use in Europe and is listed as a stabilizer, thickener, and gelling agent for the manufacturing of canned pet food. Fortunately most of the gums being texturizing components are used at very low concentrations, for example, refined and food-grade CT gum, which contains <10 ppm of chryso-phanic acid is permitted at a limit of 0.5% in pet food. Thus, there should be hardly any significant amount of chrysopanic acid in a food containing a CT gum additive.

REFERENCES

1. Arora, S. K., Ed., Chemistry and Biochemistry of Legumes, Oxford & IBH Publishing Co., New Delhi, 1982.
2. Whistler, R. L. and BeMiller, J. N., eds., Industrial Gums, 3rd ed., Academic Press, New York, 1993.
3. Anderson, E., Ind. Eng. Chem., 41 (1949): 1887–1890.
4. Cui, S. W., Polysaccharide Gums from Agricultural Products: Processing, Structure, and Functionality, Technomic, Lancaster, PA, 2001.
5. Kapoor, V. P., Sent, A. K., and Farooqi, M. I. H., Ind. J. Chem. (1989): 928–933.
6. Dea, L. C. M. and Morrison, A., Adv. Chem. Biochem. Carbohyd., 31 (1975): 241–310.
7. Hui, F. P. A., and Neukom, H., Tappi, 47 (1965): 39–42.
8. Burton, H., Chapman, H. R., and Williams, D. J. R., Proc. Int. Dairy 16th Cong., 3 (1962): 82.
9. Pederson, J. K., Cereal Sci. Today, 19 (1974): 476.
10. Liener, I. E., In Chemistry and Biochemistry of Legumes, S. K. Aora, ed., Oxford &IBH Publishing Co., New Delhi, 1982, 217–258.
11. Cassia gum, Wikipedia, September 2008.
12. Dea, I. C. M., McKinnon, A. A., and Rees, D. A., J. Mol. Bio., 68 (1972): 391–195.
13. Hui, P.A., Thesis, Zurich University, Zurich, 1961.
14. New Vistas in Pulse Production, Indian Agricultural Research Institute, New Delhi, 1971.
15. Whistler, R. L. and Hymowitz, T., Guar: Agronomy, Production, Industrial Use, and Nutrition, Purdue University Press, West Lafayette, IN, 1979.

12 Miscellaneous, Less Common Galactomannans and Glucomannans

12.1 INTRODUCTION

It is a well-recognized fact that the availability of galactomannans or the polysaccharide gums, which are derived from perennial trees and shrubs, is going to be limited. Whereas the cultivation of galactomannan-producing annual crops can be increased when such a need arises, trees cannot be grown seasonally to produce more gums. The land occupancy by full-grown trees is more or less permanent. There have been occasional, sudden spurts in the world's demand of tree products, including the gums they produce. Further, this has created a need to explore alternate trees as the sources for producing industrial gums.[1]

It has been a general observation that the legume galactomannan polysaccharides derived from trees, for example, locust bean gum (carob gum or LBG), have a lower percentage (~20%) of galactose sugar when compared with the higher percentage of galactose derived from annual agricultural legume crops, for example, guar gum (~33%) and fenugreek gum (~48%). Galactomannans with a lower percentage of galactose have poor water solubility and exhibit a mixed sol-gel rheology. Hydrocolloids with such functional properties, that is, a mixed sol-gel rheology, are an ever-increasing demand in the food industry.[2]

Besides LBG, another gum with a low percentage (~25%) of galactose is obtained from the tara shrub, which is a perennial plant, having its origin in the Andes mountainous range of Peru in South America. Tara gum was commercialized during the mid-1990s. In spite of its limited availability and higher market price, tara gum has been considered to be a reasonably good substitute for LBG in the food industry.

Cassia fistula is a tree of the South Asian region, which produces a galactomannan polysaccharide somewhat similar to LBG and tara gum.

12.2 *CASSIA FISTULA* TREE[3]

Cassia fistula (Hindi name is *amalatas*) is widely grown as an ornamental tree in tropical and subtropical countries in Asia including India, Pakistan, Burma, Ceylon,

FIGURE 12.1 *Cassia fistula* tree in full bloom.

Thailand, Indonesia, and Malaysia. The tree blooms in late spring and during the early summer, that is, in April–June in the northern hemisphere and November–January in the southern hemisphere. Flowering of this tree is profuse, with the tree getting so much coverage with yellow flowers that almost no green leaves are visible (Figure 12.1). It is also known as golden shower or the Indian labunum. Growth of this tree is best in the summer sun and on well-drained soil. The tree is drought resistant and tolerates saline soils. Being a tropical tree it is damaged by repeating spells of freezing weather.

This fast growing tree can have a height of 10–20 meters, with many spreading branches toward its summit. Its leaves are 15–60 cm long and pinnate. The leaves are deciduous and semievergreen. The tree bears large flowers of 4–7 cm sizes in diameter, which are fragrant and bright yellow in color, and these are borne on long, slender, and smooth pedicels. The tree bears fruits (pods), which are dark, blackish-brown and woody in color. These pods are cylindrical in shape, 30–50 cm in length, and about 2.5 cm in diameter. The pods are filled with a viscid, reddish-black, sweet pulp, and it is divided into many cells by hard, transverse phragmata. Numerous matured, long, blackish pods can be seen hanging on the branches of the tree, which ultimately fall down on the ground. Each cell in the pod contains one oval-shaped glossy and somewhat flattened seed. On shaking a dried pod, the rattling sound of the seeds inside can be heard.

12.3 *CASSIA FISTULA* SEED GUM: NEED TO EXPLORE PRODUCTION AND COMMERCIALIZATION[4]

Cassia fistula is considered to be a medicinal and ornamental plant as well as a shade tree. Seeds of this tree are widely available and they are underutilized.[3] Not much attempt seems to have been made to use them for extraction of their galactomannan gum.

Preliminary investigations related to *Cassia fistula* gum, which happens to be a tree-seed derived galactomannan polysaccharide, were carried out at the National Botanical Research Institute[3] (NBRI, which is one of the Council of Scientific and Industrial Research Institutes in Lucknow, India). It has been reported on the basis of these studies at NBRI that this gum holds a strong potential to become a good substitute for LBG and tara gum. There is a need to further investigate its safety as a food hydrocolloid additive.

Some investigations on the polysaccharide gum from *Cassia fistula* seeds have also been reported from Australia and Thailand, but most of the work has been confined to the lab. The whole seed powder of *Cassia fistula* has also been used in some ayurvedic drug formulations.

12.4 EXTRACTION OF *CASSIA FISTULA* SEED GUM: DRY GRINDING AND WET EXTRACTION METHODS

There are many reported methods of galactomannan extraction from different legume seeds and these can also be used for *Cassia fistula* seed. Commercially feasible and cheaper methods of gum extraction are those based on dry milling, sieving, and sifting processes. Such methods may involve dry heating of the seed or its split to loosen its hull. This is followed by differential and course grinding and sieving to separate the hard and mostly intact endosperm halves or the splits from the easily powdered germ and hull portions. Hammer milling of thus separated and dehusked seed endosperm generally gives an industrial-grade gum powder of desired purity and fineness.[2]

In the lab, the solvent extraction process (wet method) of galactomannan separation from whole seed powder has frequently been used. These methods involve aqueous or aqueous ethanol (solvent) extraction methods, which are generally used in labs. In these methods, the galactomannan portion from the whole seed powder is dissolved in hot water, followed by cooling to or below room temperature, when most of the proteins and cellulosic material remain insoluble. These insoluble constituents are centrifuged or filtered off. From the clarified filtrate, containing the aqueous gum solution precipitation of the gum is carried by the addition of a water-soluble organic solvent (alcohol or acetone).

By such wet extraction methods, the yield of galactomannan gum from *Cassia fistula* seed has been reported to be about 24%, whereas the actual gum content of the seed is more than 30%. Lower yield by solvent extraction methods may be due to incomplete dissolution of the gum, particularly in case of those galactomannans that have lower (25% or less) galactose content and consequently low water solubility. Another factor that can result in the lowering of yield is the presence of galactomannan hydrolyzing enzymes, which are generally present in the whole seed powder. These can degrade the gum to a certain extent and thus lower the gum yield. For further purification of the gum, the galactomannan polysaccharide is redissolved in water and it is reprecipitated with ethanol to give a purer product.[5,6]

From an economical standpoint, these wet extraction methods are less preferred on a commercial scale because of the large increase in processing costs. Until now industrial-scale extraction of *Cassia fistula* gum has not been done anywhere in the world and hence it is not being marketed on any commercial scale.

Seeds of *Cassia fistula* are much larger compared to those of guar and fenugreek and its husk is less pigmented compared to that of carob seed. The endosperm split of *Cassia fistula* seed is also less colored, compared to that of carob split. Hence, it should be possible to commercially produce a good quality of food gum from *Cassia fistula* seed. Extraction of the gum yield can certainly be increased in a differential dry grinding–sieving process.

12.5   FUNCTIONAL PROPERTIES *OF CASSIA FISTULA* GUM[2,7]

In its functional properties, *Cassia fistula* seed galactomannan gum has been found to be comparable to tara gum and LBG. The mannose-to-galactose ratio (M:G) in both tara gum and in *Cassia fistula* gum is close to 3:1, which is intermediate between that for guar gum (2:1) and LBG (4:1). In physical and functional properties, both tara gum and *Cassia fistula* gum are expected to behave like a blend of LBG and guar gum. Further, the blends of either *Cassia fistula* or tara gum with xanthan gum should have a rheology suitable as a food additive and it can open new prospects for their use in food. So far *Cassia fistula* gum has not found any favor for commercialization.

Samples of the *Cassia fistula* gum prepared in the lab are generally a whitish powder, containing about 10% moisture, some protein (3%–5%), and nearly 85% galactomannan polysaccharide. For other galactomannans (e.g., LBG and tara gum), the seed coats are much darker when compared to that of *Cassia fistula* seed. Due to some specks of seed coat contaminating the gum, the resulting gum products from locust bean and *Cassia tora* are, generally, somewhat reddish-brown in color. It may be expected that the gum from lighter seeds of *Cassia fistula* when commercially prepared should be whiter in color compared to other tree-seed-derived gum. Much cannot be said at this stage about other functional properties, for example, solubility and viscosity (rheological behavior), of this less explored gum.

12.6   STRUCTURAL FEATURES OF *CASSIA FISTULA* GALACTOMANNAN[3,8]

Structural studies of *Cassia fistula* gum by Dr. V.P. Kapoor were carried out at the National Botanical Research Institute.[3] These have revealed the usual structural features of legume galactomannans for this polysaccharide.

Mucilaginous endosperm powder of *Cassia fistula* seeds is mainly composed of a galactomannan, which is a neutral polysaccharide.[7] In general the legume seed galactomannan molecules are composed of a linear backbone of β(1→4)-linked D-mannopyranose units, with α(1→6)-linked side chains (graft) of a single D-galactopyranose unit. These structural features are also present in the *Cassia fistula* galactomannan molecule. The M:G ratio in *Cassia fistula* gum has been found to be close to 3:1.[3]

As is well known, the ratio of D-mannose and D-galactose in galactomannan polysaccharides varies in plant seeds from one species to the other species. Also the ratio (M:G) varies, within the aqueous solubility based fractions of the polysaccharide from the same species. This is true of *Cassia fistula* gum also. The M:G ratio in polysaccharide samples from plants of the same species but from different geographical locations has been reported to differ. It has been suggested that this difference may arise due to variable environmental factors. This is also true of *Cassia fistula* gum.

An important functional property of all the galactomannan gums is their high water-binding capacity and the formation of viscous aqueous solutions at a relatively low concentration of the gum. This is also true of *Cassia fistula* gum. In food applications, the galactomannan gums are mainly used for their moisturizing, efficient thickening, binary gelling, and stabilizing effects. These functions are likely to be satisfactorily done by *Cassia fistula* gum.

For tara gum and *Cassia fistula* gum, the distribution of galactose grafts on their mannan backbones has not so far been studied. Binary gelling of xanthan gum and agar with tara gum and *Cassia fistula* gum suggests that the galactose stubs in both these polysaccharides are likely to be present in continuous blocks of side-chain grafts, which are separated by bare mannose blocks. Thus, the distribution of galactose grafts in *Cassia fistula* gum is likely to be similar to that in LBG and tara gum.

Both tara and *Cassia fistula* gums show weak gelling properties, which are comparable to that in LBG (M:G = 4:1). If produced commercially, *Cassia fistula* gum can thus find applications in the food industry as a substitute for LBG. After toxicity studies, *Cassia fistula* gum could be approved as a food additive. In many food applications, it can act as a suitable stabilizer and useful in mixed gel systems.

12.7   WHY EXPLORE AND COMMERCIALIZE *CASSIA FISTULA* GUM?[8]

The prospect of *Cassia fistula* gum, if it is commercialized, is difficult to predict. However, looking to its widely spread and existing plantations, the production amount of this gum in India alone could be comparable to or even more than that of tara gum (1000–1500 tons per annum). Just like tara gum, the product gum of this tree is likely to be used exclusively by the food industry as a thickener and stabilizer, where it can fetch much higher prices compared to that of guar gum. At present India does not produce any galactomannan gum derived from trees having lower a percentage (M:G in the range 3:1) of galactose. If commercialized, *Cassia fistula* gum could be the first one of this type of galactomannan to be produced in India.

The tara pod was earlier processed only for its tannins in Peru. After breaking its pod for extracting tannins, tara seeds were not in use for gum extraction. These were later exported to Europe (Italy and Switzerland) where its gum extraction was started in facilities using technology similar to that of LBG manufacturing.

Tannins are also likely to be present in the dark, outer skin of *Cassia fistula* pods and this needs be investigated. Fortunately galactomannan extraction technology from guar seed, which is based on differential grinding, exists in India where it is practiced on a much larger scale than anywhere else in the world. If *Cassia fistula* galactomannan gum and tannin extraction is to be initiated in India, an additional amount of seeds can even be procured from neighboring Burma, Malaysia, and Indonesia.

The current manufacturers of LBG and tara gum are powerful, multinational, hydrocolloid companies of Europe, including Cesalpinia S.p.A. and Nutragum S.p.A. of Italy and Unipectin AG in Switzerland. Compared to the processing of *Cassia fistula* seed for its gum, the processing of LBG and tara gum as followed in these companies has been more problematic particularly due to poor dehusking of the seed and difficulties in powdering of the hard endosperm of carob and tara seed endosperm.[9]

Broken pods of *Cassia fistula*, just like those of the tara pod, are likely to be rich in tannins and the recovery and commercial evaluation of these two products (gum and tannins) from its pod can result in a good value addition. Processing problems for extraction of *Cassia fistula* gum from its seed is not likely to be a problem in India because of the current expertise (machines and processing technology) available with the guar gum industry in India.

Presently, India is importing a sizable quantity of LBG (binary gelling gum) for use in its well-developed food industry. The gum from a new source, *Cassia fistula,* can replace imported LBG and thus save much-needed foreign exchange for the country. If more of this new gum is produced, it can be exported. World over, there is great demand for gums with functional properties similar to those of LBG.

It has been relatively recent (early 1990s) that tara gum production was commercialized and whole of this gum, now being produced has found a market in the food industry. It is of interest to note that exploration is being carried out to find a substitute for LBG. As a specific example, galactomannan polysaccharide from another cassia species, *Cassia brewsteri* seed, has also been investigated in Australia as yet another galactomannan gum as a substitute for LBG.

12.8 WHAT CAN BE THE POSSIBLE USES OF *CASSIA FISTULA* GUM?[10]

The food industry is likely to be the main user of *Cassia fistula* galactomannan gum, which has LBG-like functionalities. This gum has also been found suitable as a binder for pharmaceutical tablets.[11] Compared to the gums from agricultural crops (guar and fenugreek) of Indian origin, the cost of production of *Cassia fistula* gum is likely to be high, due to the labor cost involved in collection of its seed and higher processing costs. A favorable point should be the hardness of its endosperm split, which is generally harder for legume splits having a lower percentage of galactose. If commercial production of *Cassia fistula* gum is undertaken, the preferred extraction method shall be differential dry grinding of the seed. A less developed country like Peru has found tara gum as a source of revenue and foreign exchange. *Cassia fistula* may prove the same for India.[12]

Blends of either *Cassia fistula* or tara gum with xanthan gum (a microbial polysaccharide) has a rheology that could open up new vistas for their use in the food industry. Tara gum is currently produced in a small amount (1000–1500 tons per year), which is primarily used in the food industry as a thickener and stabilizer. It is used in dessert gels, seafood, meat, ice cream, and yogurt. It has improved rheology, water binding, and emulsifying properties. Patents related to tara gum claim it to be a water-soluble polymer, which has applications comparable to guar or LBG, but the price structure of tara gum does not justify its functional application other than for food.

Distribution of galactose grafts on tara gum backbone has not been studied in full detail, but its binary gelling of xanthan (microbial origin) gum and agar (seaweed gum) suggests that galactose stubs are likely to be present in blocks. This is also true of *Cassi fistula* gum, which has a weak self-gelling property and also forms mixed gels with xanthan. There is a need for toxicity studies of *Cassia fistula* gum before it can be used as a food additive.

12.9   GALACTOMANNAN FROM PODS OF *PROSOPIS JULIFLORA*

Prosopis juliflora (which is called *vilayati babool* in India) was introduced into India over a hundred years ago from Central America. It has now spread all over the country, particularly in the northwestern, semiarid regions and the wastelands. The tree is only being used as a source of cheap firewood, while its matured and dried yellow pods are used as fodder for cattle, particularly for sheep and goats in the states of Rajasthan, Gujarat, and Haryana. This tree also produces an exude gum, which has not been exploited on a commercial scale because of very low yield.

The peculiarity of this tree lies in the fact that when growing in a crowded space, it grows only as a shrub, but with isolating a plant and pruning, it can grow into a large tree. It is only the full-grown trees that bear pods. Initially the tree bears 1216 cm long green pods, which turn yellow on maturing, then drying and falls off the tree when shaken. The dried and flattened yellow pods of this tree consist of an outer skin and a layer of dried pulp, rich in sugars and underneath lays the protein-rich seed. Since the pods of this tree cannot be fully digested by cattle unless it is broken and ground, small-scale processing of pod into powder for cattle feed has now been started in parts of Rajasthan and Gujarat states in India.

Endosperm of the dicotyledonous seed of this tree is composed of a galactomannan. Studies by this author have revealed that just like many legume tree seeds, the endosperm portion of *Prosopis juliflora* seeds contains a galactomannan, which appears similar to guar gum but is of a much lower viscosity. Not much information is currently available about the chemical structure, M:G ratio, and other properties of this galactomannan. Since the tree produces an exude gum, as well as a seed galactomannan, there has been some confusion about the nature of these two polysaccharides. It should be clearly understood that the seed gum and not the exude polysaccharide is a galactomannan.[13]

Looking to the currently available and commercially produced guar galactomannan in India, *Prosopis juliflora* seed is not likely to be exploited as a source of industrial galactomannan polysaccharide in the near future.

12.10 GLUCOMANNAN POLYSACCHARIDES[14]

Glucomannans are yet another group of high molecular weight polysaccharide gums that have certain functional properties and uses similar to those of galactomannans. The polysaccharide backbone in these gums is composed of blocks of $\beta(1\rightarrow4)$-linked mannopyranose sugars having blocks of interposed $\beta(1\rightarrow4)$-linked glucopyranose. The M:G ratio in konjak polysaccharide is approximately 1.6:1. Glucomannans are mainly linear-chain polymers. There can also a small amount of branching due to single glucopyranose groups present as grafts, as is the case of konjak polysaccharide. Such branching has been reported to an extent of about 8% of the total monosaccharide groups present in konjak glucomannan. These single sugar grafts are $\beta(1\rightarrow6)$-linked glucopyranosyl groups on the linear backbone. Some sugar groups of the polysaccharide chain are di- or even tri-acetylated in 2, 3, and 6 positions of the backbone hexoses. Water solubility of these glucomannans has been attributed to the presence of these short side chains, comprised of acetyl groups, and single glucose grafts. These short grafts on the linear backbone of the polysaccharide prevent interchain hydrogen bonding and do not permit aggregation of polymer molecules into a insoluble mass.

12.11 KONJAK GLUCOMANNAN

Konjak is a large tuber plant that is extensively grown in India, China, and most of the East Asian countries. Konjak flour or its isolated glucomannan polysaccharide is used for making food gels, where it is used as a cheaper substitute for gelatin. Being nondigestible in the human system, konjak glucomannan functions as a soluble dietary fiber. It contributes almost no calories when used in food and thus it is commonly used as a diet food.

Konjak plant produces large tubers that are 20–30 cm in diameter and can weigh up to 2–5 kg. The tuber contains a glucomannan as its major polysaccharide, forming up to 40% of its dry mass. This unique glucomannan is a water-soluble dietary fiber. Konjak glucomannan is a very high molecular weight (200,000 to 2,000,000 dalton) polysaccharide. Its molecule is built up of a linear chain, which consists of a larger block of $\beta(1\rightarrow4)$-linked mannopyranose sugars and having interposed $\beta(1\rightarrow4)$-linked glucopyranose groups.[15]

Konjak glucomannan has a property of swelling, by absorbing more than a hundred times its own weight of water.

In most instances, researchers, who were to elucidate the structure of this polysaccharide and its impact on human health, have made claims that incorporation of konjak glucomannan as a dietary fiber in human food:

1. Promotes loss of body weight
2. Lowers blood cholesterol and triglyceride levels

3. Reduces absorption of fats from the digestive tract
4. Promotes control of blood-sugar levels
5. Promotes feeling of fullness or satiety
6. Causes improvement of diet for diabetic persons

12.12 USES AND APPLICATIONS OF KONJAK GLUCOMANNAN

Konjak glucomannan is a water-soluble polysaccharide gum, which acts as a soluble dietary fiber. It is not metabolized in the human digestive system. Although the U.S. Food and Drug Administration has not yet approved it as a nutritional supplement, many clinical trials have established that konjak gum cures constipation, obesity, and controls blood sugar and lipid level, including triglycerides and cholesterol. Many dietary food preparations that contain konjak glucomannan as a dietary fiber are now available in some Western countries. There is however little doubt that besides certain definite health benefits, there is also a lot of exaggeration about the beneficial action of most of these glucomannan polysaccharides.

12.13 ALOE VERA GLUCOMANNAN[16]

Aloe vera is another plant, whose leaves have a gelled form of glucomannan polysaccharides. Because of a very low percentage of glucose units in it, compared to that of mannose, and high acetyl content, these glucomannans are also referred to as *acemannan* or *acetylated mannan*. The amount of glucomannans in aloe vera leave gel extract is very low, less than 1.0 %; the rest being water and some other minor constituents. Extraction of pure (dry and solid) aloe vera glucomannan polysaccharide has only been carried out in laboratories and that too for its chemical structure elucidation.

Aloe leaf juice containing glucomannan has been extensively used in medicinal and cosmetic products. On a commercial scale, aloe vera glucomannan is sold as a fluidextract of leaves or as binary gels with carrageen or Carbapol (trade name of an acrylic polymer).

12.14 ALOE VERA GLUCOMANNAN, GEL OR FLUID

On transverse cutting of a thick aloe vera leaf, one comes across a semifluid gel and hence it is commonly referred to as native aloe vera gel. On shearing during its extraction, this semifluid gel, it permanently transformed into a fluid. The so-called, commercially produced aloe gels are in fact binary gels, where aloe vera glucomannan is only a minor component of the gel.

One unique property of aloe vera glucomannan is its very high affinity for moisture, and its property to prevent loss of moisture, which makes it an extremely useful and active component of cosmetics even when it is present at a very low concentration. Why does aloe vera glucomannan have this extreme power to prevent moisture loss? Structural studies of aloe vera glucomannan have shown it to have a high degree of acetyl substitution. It is also established from chemical structural studies of aloe vera polysaccharide that even at a high degree of substitution (DS) of acetyl groups, the substitution of acetyl groups in aloe vera polysaccharide is highly non-uniform. The acetyl substituents are localized only into a few hexoses. Acetyl substituted hexoses in aloe vera polysaccharide are generally di- or trisubstituted, which makes those hexoses hydrophobic.

Those segments of aloe vera polysaccharide backbone, which are substituted by two or three acetyl groups, has acquired enough hydrophobic character to get segregated from the rest of the hydrophilic segment of its polymer chain. Such segregation is entropically favored, and thermodynamically this is possible in large and linear macromolecules, such as those of aloe vera glucomannan. Incompatibility between hydrophobic and hydrophilic segments of this linear chain polymer results in a microphase separation. A strongly hydrated, hydrophilic phase is entrapped in an outer layer of hydrophobic microphase, resulting in gelling. Shearing disturbs the outer hydrophobic phase, resulting in synerosis and thixotropic rheology.

In aloe vera leaves, glucomannan (which has no tertiary structure) is present at a concentration that is less than 1.0% concentration in an aqueous media. No polysaccharide is known to produce a gel at such low concentration. The following could be the reason for the gelling.

- Gelling arises due to the formation of a complex between the glucomannan polysaccharide and a microcrystalline cellulose-type polysaccharide, which is also present along with it in the aloe vera leaf. Such a complex mixture of two polysaccharides has been termed a *hydrocolloid alloy.*
- According to the author, gelling of aqueous aloe vera glucomannan is due to segregation of the hydrophobic and hydrophilic segments of its linear molecular chains. Due to microphase separation, a thin hydrophobic film encases the bulk of hydrophilic fluid, giving it a gel-like appearance.

REFERENCES

1. Cui, S. W., Polysaccharide Gums from Agricultural Products: Processing, Structure and Functionality, Technomic, Lancaster, PA, 2001.
2. Srivastava, M. and Kapoor, V. P., Seed galactomannans: An overview, Chem. Biodiv., 2 (2005): 295–317.
3. Chaubey, M. and Kapoor, V. P., Structure of a galactomannan from the seeds of Cassia angustifolia Vahl, Carbohyd. Res., 332 (2001): 439–444.
4. Cunningham, D. C. and Walsh, K. B., Galactomannan content and composition of Cassia seed, Austral. J. Exper. Agr., 42 (1985): 1081–1086.
5. Thunyawart, J., Karagred, K., and Sittkijyothin, W., Galactomannan extraction from Cassia fistula seed, www.turpif.or.th/project_reward/project_file/2550_2008-06-30_F127_R50B10001_Complete.pdf.
6. Kapoor, V. P., Chanzy, H., and Taravel, F. R., X-ray diffraction studies on some seed galactomannans from India, Carbohyd. Polym., 27 (1995): 229–233.
7. Duke, J. A., Handbook on Energy Crops, Purdue University, Microsoft Encarta, 1998.
8. Buckeridge, M. S., Panegassi, V. R., Rocha, D. C., and Dietrich, S. M. C., Seed galactomannan in the classification and evolution of the leguminosae, Phytochemistry, 38 (1994): 871–875.

9. Lazaridou, A., Biliaderis, C. G., Izydorczyk, M. S., Structural characteristics and rheological properties of locust bean galactomannans: A comparison of samples from different carob tree populations. J. Sci. Food Agri., 81 (2000): 68–75.

10. Mathur, V. and Mathur, N. K., Fenugreek and lesser known Galactomannan polysaccharides: Scope for development, J. Sci. Indust. Res., 64 (2005): 475–481.

11. Andrade, C. T., Azero, E. G., Luciano, L., and Gonalves, M. P., Solution properties of the galactomannans extracted from the seeds of Caesalpinia pulcherrima and Cassia javanica: Comparison with locust bean gum., Int. J. Biol. Macromol., 26 (1999): 181–185.

12. Monif, T., Malhotra, A. K., Kapoor, V. P., Cassia fistula seed galactomannan: potential binding agent for pharmaceutical formulation, Ind. J. Pharm. Sci., 54 (1992): 234–240.

13. Mathur, N. K., Unpublished observations.

14. Bittiger, H. and Husemann, E., Polym. Sci. Part B, 10 (1972): 367–371.

15. Kato, K. and Matsuda, K., Agri. Biol. Chem. (Tokyo), 33 (1969): 1446–1453.

16. Gowda, D. C., NeelSiddaiah, B., and Anjaneyalu, Y. V., Carbohyd. Res., 72 (1979): 201.

Index